...R VOLES

Rob Strachan
May 97

# WATER VOLES

written and illustrated by
Rob Strachan

**Whittet Books**

First published 1997

Whittet Books Ltd, 18 Anley Road, London W14 OBY



British Library Cataloguing in Publication Data. A catalogue record for this book is available from the British Library.

ISBN 1 873580 33 9

Printed and bound in Great Britain by
Biddles Limited, Guildford and King's Lynn

# Contents

# Acknowledgments

Perhaps my first encounters with the cast of *The Wind in the Willows* came from my regular childhood summer holiday trips to a house-boat based on the Norfolk Broads. I distinctly remember the early morning expeditions by rowing boat I would make with my brothers to a secret place we called 'Toad Island', alive with frogs and toads; a half eaten fish and footprints would be tell-tale signs of an otter; there were molehills (and so presumably moles); seemingly, everywhere we looked a water vole could be seen sitting on his haunches munching a stem of reed. I would like to thank my parents for introducing me to the joys of this natural world, with the chance to grow up in 'the countryside' and tolerating my fascination with animals weird and wonderful which often found their way back to my bedroom!

The chance really to get to know the water vole came in 1989, when The Vincent Wildlife Trust asked me to conduct the first ever systematic survey of the species in England, Scotland and Wales. Living in a camper van, taking two years to complete it, I scoured the waterways, from Land's End to John O'Groats, visiting 2,970 sites and walking 1,200 miles of riverbank searching for the signs of water voles. I am indebted to Vincent Weir and Don Jefferies for their foresight that such a survey needed to be done and their confidence that I would complete it on time. The Trust is currently re-surveying all the water vole sites and I am very grateful to be given the opportunity to join in the project again. Thanks are also due to the staff at the VWT office who make the whole operation run so smoothly.

Since 1994, I have been working for the Wildlife Conservation Research Unit at Oxford University, studying the water vole in much more detail, researching the factors implicated in the vole's decline and attempting to find practical answers for its conservation. The teamwork and companionship of the 'WildCRU' has been brilliant; I realise now what had been missing from my previous ten years of being 'on the road'. Special thanks are due to the head of the Unit, David Macdonald, for his warm welcome, infectious enthusiasm and role as mentor to my work. I have really enjoyed the humour and camaraderie of my fellow vole and mink researchers, Guillermo Barreto and Nobuyuki Yamaguchi, and all our dedicated radio-tracking field assistants (I remember those nights when it was so bitterly cold that the radio-collared mink decided to stay in its nice snug, warm den inside the hollow tree all night – and you stayed on the opposite exposed bank, in snow and a wind blowing all the way from Siberia, just to prove it!).

The vision of Alistair Driver who put the water vole on the agenda with the Environment Agency must also be acknowledged; we now have the opportunity of putting the clock back, restoring and enhancing our waterway habitats and securing a future for Ratty. I must also thank all those I have not mentioned but whose work and anecdotal stories have contributed to this book and to Annabel Whittet for relenting to my badgering to write it.

Finally: Thank you, Jackie, this is for you.

# Preface

As I write this within view of the dreaming spires of Oxford, I am reminded not only that the author of *The Wind in the Willows*, Kenneth Grahame, is buried in one of Oxford's cemeteries, but that one hundred years ago he went to school here and would have spent many of his long summer days of childhood along the banks of the River Thames. Perhaps the common sight of water voles chugging about across the surface of the river, felling reed stems like mini-lumberjacks, diving under water with a familiar 'plop' or darting into one of the myriad of bankside holes gave him the inspiration for one of the classic children's books. He captured some of the magic of life by the river, quintessentially all that is England, in a book that was first published in 1908 but is as popular now as it ever was.

Since that time there have been big changes in the British countryside: rivers have been engineered, dredged and realigned, made to conform to the straight lines of man's neat and tidy mind. Wetlands have been drained, pastures have been ploughed, hedgerows removed to make way for the growing of crops in large fields that can be harvested by increasingly bigger and better machines (much to the horror of Badger and Mole). Roads criss-cross the countryside, increasingly straighter, wider, faster (much to the delight of Mr Toad – poop, poop!) and urban development follows in its wake. The land has changed and is still changing – 'improved' and 'modernised', in our progress through the twentieth century. The wildlife too has changed and is still changing in its distribution and abundance.

A few species, like the urban fox, the rabbit and brown rat have been quick to adapt to the new environment and are doing very well but others are finding it increasingly difficult to survive. Some are in need of direct action to help save them from the threat of extinction. The water vole, best loved as Ratty, from *The Wind in the Willows*, is one such animal.

This is his story.

# Ratty

*As he sat on the grass and looked across the river, a dark hole in the bank opposite, just above the water's edge, caught his eye, and dreamily he fell to considering what a nice snug dwelling-place it would make for an animal with few wants and fond of a bijou riverside residence, above flood level and remote from noise and dust. As he gazed, something bright and small seemed to twinkle down in the heart of it, vanished, then twinkled once more like a tiny star. But it could hardly be a star in such an unlikely situation; and it was too glittering and small for a glow-worm. Then, as he looked, it winked at him, and so declared itself to be an eye; and a small face began gradually to grow up around it, like a frame around a picture.*

*A brown little face, with whiskers.*

*A grave round face, with the same twinkle in its eye that had first attracted his notice.*

*Small neat ears and thick silky hair.*

*It was the Water Rat !*

And so we are introduced to Ratty, the stoical star of Kenneth Grahame's *The Wind in the Willows*. However, Ratty was not a rat at all but a vole! This beautifully illustrates our confusion over this animal.

To set the record straight I had better go back to the beginning, and unravel the various scientific and local names that have been given to our Ratty. Back in 1693 an eminent English Zoologist, John Ray, first described a large rodent found in the British waterways that was characterised by webbed feet that it used for swimming and diving, and he named it the Water Rat, *Mus major aquaticus*.

The Swedish biologist Carl Linnaeus accepted this description without seeing a specimen and in 1758 reclassified it as *Mus amphibius* in his famous *Systema naturae* (the classification system where every plant and animal is given a generic and specific name). At the same time, Linnaeus also described a large burrowing field mouse that could be found swimming along the ditches of Sweden, this he named as *Mus terrestris*.

About ten years later, the much acclaimed naturalist from Selborne, Gilbert White, in a letter to Thomas Pennant wrote, 'I suspect much that there may be two species of water rats. I have discovered a rat on the banks of our little stream that is not web-footed, and yet is an excellent swimmer and diver; it answers exactly to the *Mus amphibius* of Linnaeus ... Linnaeus seems to be in a puzzle about his *Mus amphibius*, and to

doubt whether it differs from his *Mus terrestris*'.

In the 1790s Lacepede, a French zoologist examining specimens of 'water rats' found that they showed characteristics closer to field voles than to field mice or rats. He assigned the two types to the new genus *Arvicola* (which included all water voles).

In Britain, two forms were identified by differing coat colour, first by William MacGillivray who in 1832 described the slightly smaller black water vole in Scotland. This was thought not to be a new species but a subspecies as the two forms overlapped in their range.

The debate continued over the scientific names until a full review of museum specimens took place in 1912 when an American zoologist, G.S. Miller, decided that there were six European species of water vole: *Arvicola amphibius* (Britain); *A.sapidus* (Spain and southern France); *A.scherman* (Northern central Europe); *A.terrestris* (Norway and Sweden); and also *A.musigrani* and *A.italicus* from Italy.

However, with the advancement of science and the study of chromosomes to help identify similar looking species, the number of water vole species has now been lumped to just two in Europe (*A.terrestris* and *A.sapidus*) and another in North America (*A.richardsoni*), but these are divided again into a number of subspecies.

Thus since the 1970s the British water vole has become known as *Arvicola terrestris amphibius*; it is generally associated with a semi-aquatic lifestyle and at up to 320g is somewhat heavier than the continental *Arvicola terrestris sherman*. The latter may live away from fresh water, burrowing in meadows and pastures like a mole.

*Ratty from* The Wind in the Willows *(after Ernest Shepard, the original illustrator).*

Apart from mainland Britain, the water vole is found throughout Europe and Scandinavia to eastern Siberia. In the Iberian peninsula (Spain and parts of France) it overlaps with, or is replaced by, the southern water vole, *Arvicola sapidus*. The British subspecies is generally associated with waterways but the smaller continental subspecies is a dweller of meadows and can occur at huge densities of between 200 and 500 per hectare, making it a significant pest of orchards, bulb and rice fields in France, Switzerland, West Germany, former Yugoslavia and the Netherlands.

# What is a water vole ?

Commonly known as the water rat, the water vole is the largest of the British voles (the others being the bank vole, field vole and Orkney vole) weighing between 200–350g. It is often mistaken for the brown rat, *Rattus norvegicus,* as both species swim, dive and jump well. However, the latter species weighs much more at around 500g for adults. Otherwise the two species are easily distinguished by the shorter, more rounded body, blunt muzzle and short round ears almost hidden in the fur of the water vole and the pointed muzzle, larger eyes and pronounced ears of the brown rat. The tail length of the adult water vole measures around 130–135mm (around 60% of the mean head and body length), so is shorter actually and relatively than that of the brown rat (80-100% of head and body length) which is very sparsely furred and so can look long and scaly.

Although it swims and dives well, the water vole is not particularly adapted to water since it does not have webbed feet and its fur becomes water-logged with prolonged submergence. While leafing through the scientific literature on the species I came across one paper that described a water vole being placed in a large aquarium to observe its diving and swimming behaviour. After only three minutes' swimming the vole was seen to physically tire and make repeated attempts to climb out. A little while longer and its fur began to take on water and the animal began to sink! In the wild, dive times rarely exceed 20 seconds. All four feet are used for propulsion during swimming and diving, the vole appearing to swim in a jerky doggy-paddle style. Turning is accomplished by a firm push of the forefeet rather than the use of the tail as a rudder.

They are adapted to living in burrows, which are dug with the large front teeth (unlike the mole which has modified large front paws for digging). When burrowing the vole at first takes successive bites at the earth, the lips closing off the diastema behind the incisors to prevent soil being swallowed. Excavation is then progressed by pushing the soil behind it with the limbs and hind feet before the process is repeated in a stereotyped mechanical way.

### Teeth

The skull of a water vole if found is easily recognised by its high crowned cheek teeth. These are usually rootless and continue to grow throughout life. Wear on these molars reveals a complex pattern of prisms (areas of soft dentine surrounded by walls of hard enamel). These patterns are

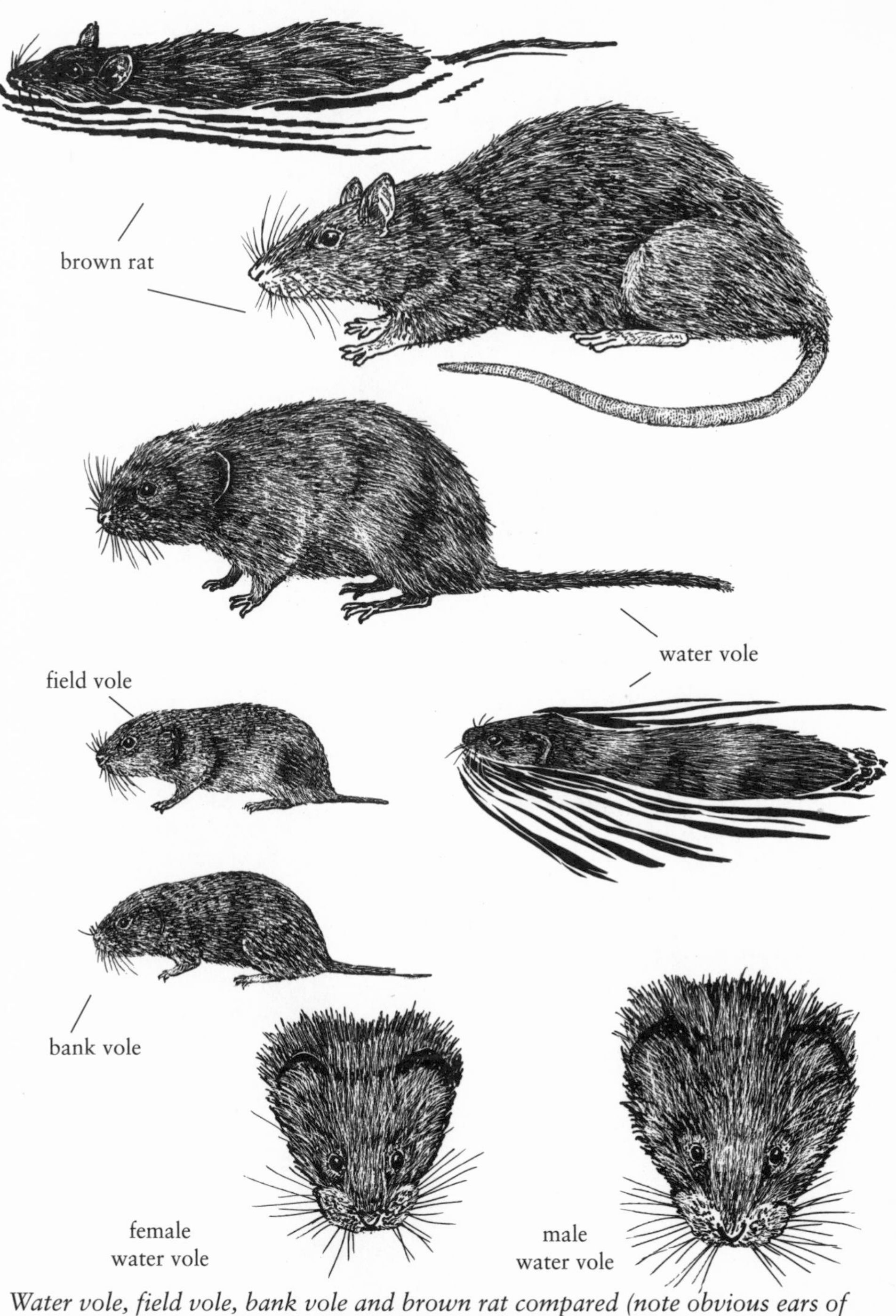


*Water vole, field vole, bank vole and brown rat compared (note obvious ears of the brown rat especially when swimming). Male water voles have a wider skull and muzzle than females.*

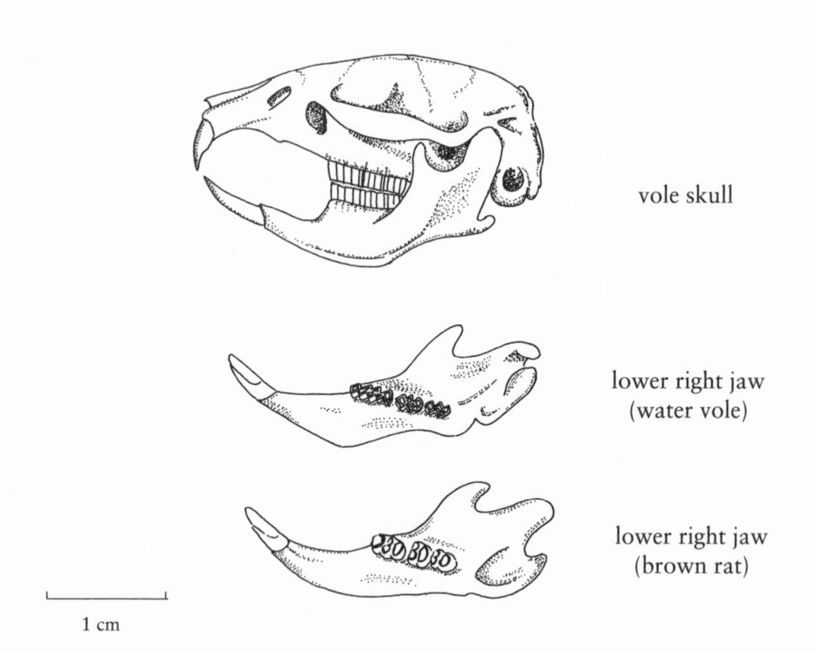


*Examining the skull and teeth of water voles will help distinguish from similar sized brown rats.*

characteristic of the species (see drawing). The bright orange incisors continue to grow all their lives and are worn down by gnawing. They have a very healthy bite too, as I know to my cost from trying to pick one up without holding it properly.

**A coat of many colours**

The tail, feet and ears are brown coloured and typically the coat colour is a rich brown tending to reddish on the back and grading to a lighter greyish-ochre on the belly.

However, in the north and north-west of Scotland the populations are predominantly melanic (pigmented black). One study revealed only 3.2% of a population near Aberdeen to be brown and a population in Wester Ross consisted of 70% melanic, 20% brown and 10% 'black and tan'. Melanic populations also occur in the Fens of East Anglia and melanistic individuals have historically been reported in virtually every county in Britain.

Other coat colours have included pure white albinos, sandy-fawn or cinnamon coloured (known as 'leucistic'), bright-buff or golden-brown (known as 'flavistic') and even pied examples, although these are exceptionally rare events. These different colour forms are genetic oddities usually lacking some or all of the skin pigments known as melanins.

However, there is a high frequency of partial albinism among most populations. This is usually observed as a white tail tip, white patches of fur on the forehead and chest or more rarely on its dorsal coat and as one or more white paws. By carefully observing your local population of water voles you will soon get to recognise the individuals with white fur patches.

Water voles undergo two moults during the course of a year, the moult line progressing from sides to back, or more rarely head to rump. The main moult is in the autumn when both juveniles and adults grow a thicker coat of long and glossy guard hairs which give the animals a darker appearance than in summer (juveniles before their first moult look particularly fluffy and less glossy than the adults). Animals that survive their first winter show a spring moult, but those surviving a second winter rarely moult a new coat and their fur appears thin and greyish.

Two types of hair are found in the water vole's coat: long silky guard hairs and short wavy under-hairs. Both insulate the water vole against heat loss and also help trap a layer of air while the animal swims and dives, giving a silvery appearance when under water. This adaptation is common among all freshwater semi-aquatic mammals, such as beaver, coypu, muskrat, mink and otter. Probably because of its small size its fur has never been prized for trade, although one Victorian naturalist with the grand name of Heneage Cocks did make a small fireside rug out of water vole skins collected from the River Thames, near Marlow.

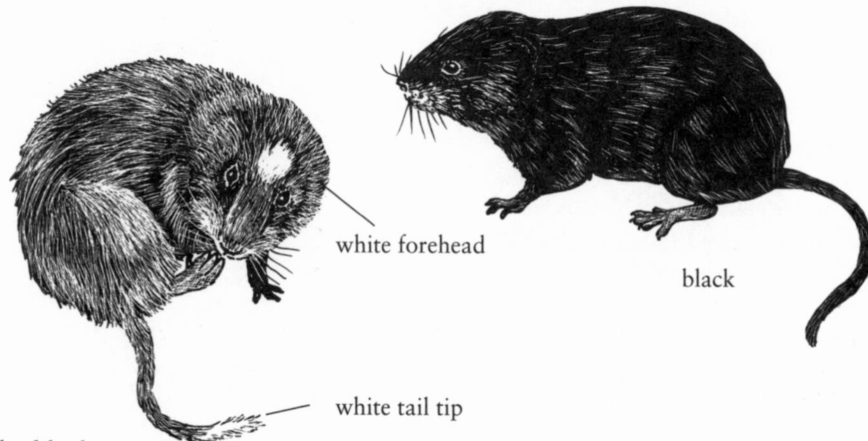


*Silky black water voles are common in Scotland while partial albinos may occur in most populations.*

# Where do they live?

*'And you really live by the river ? What a jolly life !'*

*'By it and with it and on it and in it,' said the Rat. 'It's brother and sister to me, and aunts, and company, and food and drink, and (naturally) washing. It's my world, and I don't want any other. What it hasn't got is not worth having, and what it doesn't know is not worth knowing. Whether in winter or summer, spring or autumn, it's always got its fun and its excitements.'*

The water vole once inhabited the banks of virtually every river, stream, ditch, canal, lake and pond throughout its range in Britain, favouring sites that offered thick and luxuriant swathes of waterside plants for food and shelter and a suitable bank in which to dig a series of burrows to underground nest chambers.

In Britain the best water vole populations can be found where the conditions favour a slow-flowing water course, less than 3 metres wide, around 1 metre in depth and with limited fluctuation of water levels. Permanent water is essential during periods of low flow in summer, while sites that suffer total submersion during protracted periods of winter flooding on a regular basis are untenable.

Where they are found, the bank is predominantly earth (rather than gravel or rocks) with either steps or a steep incline into which the voles can burrow and create nest chambers above the water table. Water voles may occupy inundated marshland adjacent to water courses provided there are nesting opportunities among tall tussocks of sedges, rushes and grasses. Above ground, the water vole's activity is largely confined to runs in dense vegetation within two metres of the water's edge.The amount of bankside and emergent vegetation cover is very important; the best sites offer a continuous swathe of tall and luxuriant riparian plants that provide at least 60% ground cover. Sites excessively shaded by shrubs or trees are less favoured.

Backwaters, side streams, permanent ditches and dykes and ox-bow lakes are important refuges for water voles; they may even prefer them to the main river channel.

# Watching and looking for water voles

A good way to tell if water voles are about and to see how many of them there are is to look for field signs, such as footprints, burrows and faeces. But you are also quite likely to see one, as they are among the easiest of the small mammal species to watch. They may be active during the daytime, particularly in early evening and can be observed by the patient observer sitting quietly on the riverbank.

The characteristic 'plop' of a diving water vole is usually the first sign that they are about on a river, stream, pond or canal, and that your movement disturbed one. They are very wary of any sudden looming shapes and shadows so the best bet is to move slowly and deliberately when you spot a vole (before it spots you). Crouch down or sit quietly and the water vole will accept your presence. The best places to watch are where you have found well used latrines or occupied burrows. Looking across the watercourse to the opposite bank will often give the best chance, particularly if the evening sun is shining directly onto that bank.

There is something very magical and rewarding about seeing one emerge

*Water voles dig burrows with their sharp incisors, kicking the soil out with their hind feet.*

*Look for the signs of water voles among the fringing vegetation of water courses and lake margins.*

from its burrow, slip into the water hugging close to the bank, swimming its clockwork style doggy-paddle as it patrols its home range. The sight of it then felling reed or sedge stems and swimming to a favoured feeding platform with a tasty morsel in its mouth is never to be forgotten. Try placing pieces of apple on the water's edge close to the burrows and then sit back where you have a clear view of the vole as it comes out to feed. It is even possible to get water voles used to your presence and bold individuals may be encouraged to take slices of apple from your fingers.

**Field signs**

As with many of our British mammals, water voles can also be maddeningly elusive, so in order to determine their presence you will have to become a mammal detective and look for the tell-tale evidence they leave behind.

(i) **Footprints:** Examining the soft mud at the water's edge will quickly put you on the right track. Many animals can be identified by the tracks and trails they leave and you will soon notice these rodent tracks among the larger footprints of perhaps, fox, badger, mink and even otter (my *Mammal Detective* book in this Whittet series will tell you all you need to know and a lot more besides!).

**Water vole** imprints show four toes in a star arrangment from the fore foot and five toes of the hind foot with the outer ones splayed, but often the tracks of the hind feet partially overlap those of the fore. The hind foot typically measures between 26-34 mm and is noticeably smaller than that of the **brown rat** measuring 40-45 mm (the larger, heavier rat tends to leave a deeper impression). In the water vole, the hind feet are slightly fringed with hair for swimming and show only five pads which distinguishes them from brown rats (which have six pads). Right and left tracks lie about 45mm apart and the stride averages 120mm.

Tracks are notoriously difficult to distinguish and tracks made by young rats will look identical to those made by voles.

The soft riverside mud and silt will also let you track down any mink or otters that may occur there. These are both members of the same weasel family and so have very similar features – five toes on each foot and a bounding gait when running that results in groups of all four feet registering close to one another.

The most distinctive field sign left by a **mink** is its footprints, which typically follow the soft margin at the water's edge and so are easily found. Five toes radiate from a crescent shaped central pad and can be clearly seen on soft mud, but often only four toes register on a harder surface. The claw marks are easily visible making the toe imprints pear-shaped. The larger male may give a print measuring 45mm long by 45mm wide, whereas juvenile and female prints may measure 30mm long by 25mm wide.

**Otter** tracks are much larger than those left by mink, but generally show the same arrangement of five toes in a crescent above a broad pad. Claw marks are rarely seen (unlike dog tracks) and the webbing between the toes can sometimes be seen on soft mud. The span of the footprint is about 60mm wide by 60-70mm long.

**(ii) Vole run or rat run?** Water vole runs typically occupy an area within two metres of the water's edge and take the form of low tunnels pushed through the vegetation. Pathway width is typically 5-9cm broad and may branch many times, leading to the water's edge or burrow entrances or favoured feeding areas.

Because rats are more nocturnal than water voles, their runs can be found out in the open much more (voles like to hide in vegetation cover and not cross open space, whereas the rats are using the cover of darkness). Rat runs linking their burrow entrances are very noticeable, being well used and usually bare of vegetation.

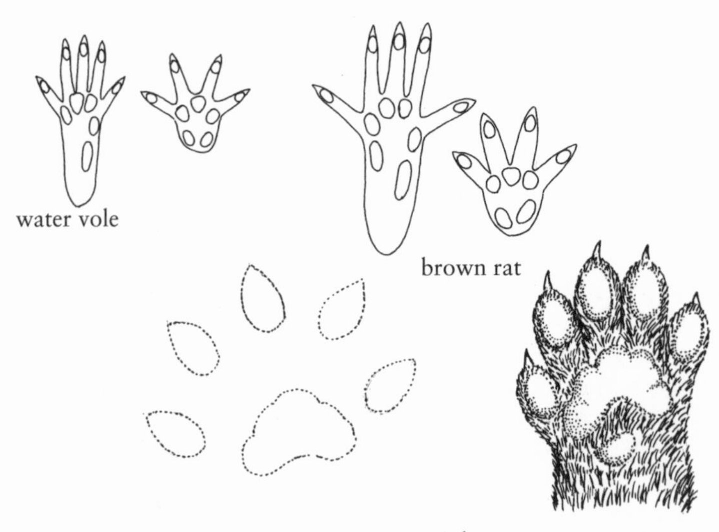


*Tracks of water vole, brown rat, otter and mink feet; both mustelids (otter and mink) have a similar bounding gait causing the track to show groups of four or two imprints. (Three-quarters of life size)*

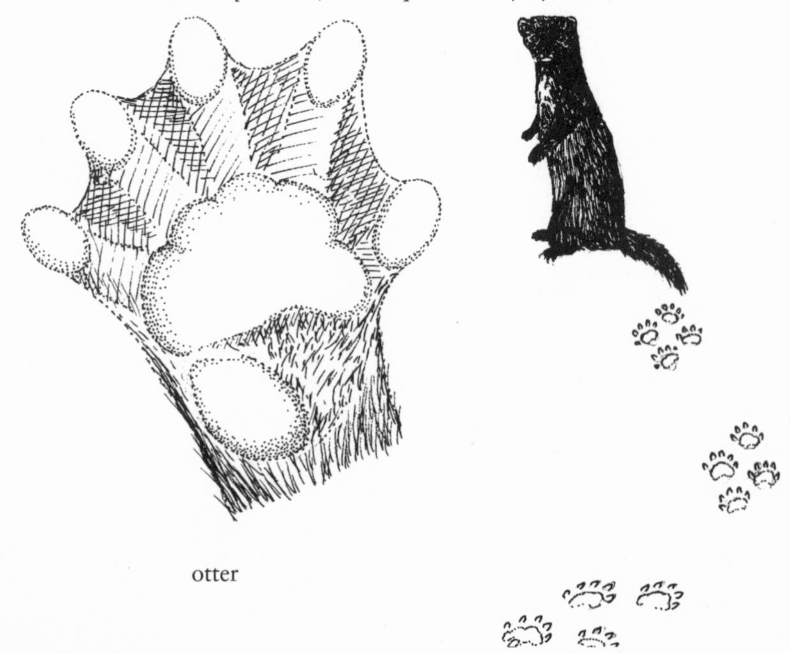

**(iii) Vole burrows and other holes in the riverbank** Searching a waterside bank will reveal many holes of various sizes. Medium sized holes of between 4 - 8cm could be made by water voles. However, beware of mole tunnels that have been revealed through the erosion of the bank. On steep earth cliffs holes made by sand martins or kingfishers may also cause confusion. Wood mouse and field vole holes are generally too small (2-3 cm across) but the burrow system of rats has to be examined carefully (rats may also take over a former water vole site).

Within a **water vole**'s home range, two types of burrow system can be found. The first is a simple short tunnel that may end in a chamber and is used as a 'bolt-hole' to escape danger or as a temporary retreat. The second forms a more complicated system of branching tunnels, multiple entrances and a number of chambers. Externally, the burrow system appears as a series of holes along the water's edge, some at or just above the water level on steep banks, some opening below the water line and others occurring within the vegetation up to three metres from the water. The hole is typically slightly wider than high, oval shaped.

Some of the entrances may look wider, due to the erosion from a rising water level (looking down the hole you will see that the tunnel contracts down to the regular 4-8cm). Spoil from the burrows is probably washed away from those entrances that open at the water's edge. Holes higher on the bank (for ventilation or escape) are probably excavated upwards from underground as no spoil can be found around them. Around these land holes, grazed 'lawns' can be found. These frequently occur when the female is nursing young and time away from the nest is kept to a minimum. The female grazes the vegetation short to the ground within easy reach of the hole; often she will not fully leave the hole, so that she can dart back should danger threaten.

Burrows dug by **rats** on the riverbank tend to be higher up the bank, away from the water's edge and grouped closely together. Each hole (6-10cm across) may show a fan-shaped spoil heap of soil outside and obvious well trodden pathways linking them up. Both rats and water voles live in colonies, but whereas the rats pack in tightly together, often sharing nests and the burrow entrances that lead to them, water voles organise themselves into a series of exclusive territories defended by each female and so string themselves out along the waterway.

**(iv) Nests** Water vole nests are usually constructed underground and so you will not normally find them, but where the water table is high and vegetation cover is dense, nests can be occasionally encountered above

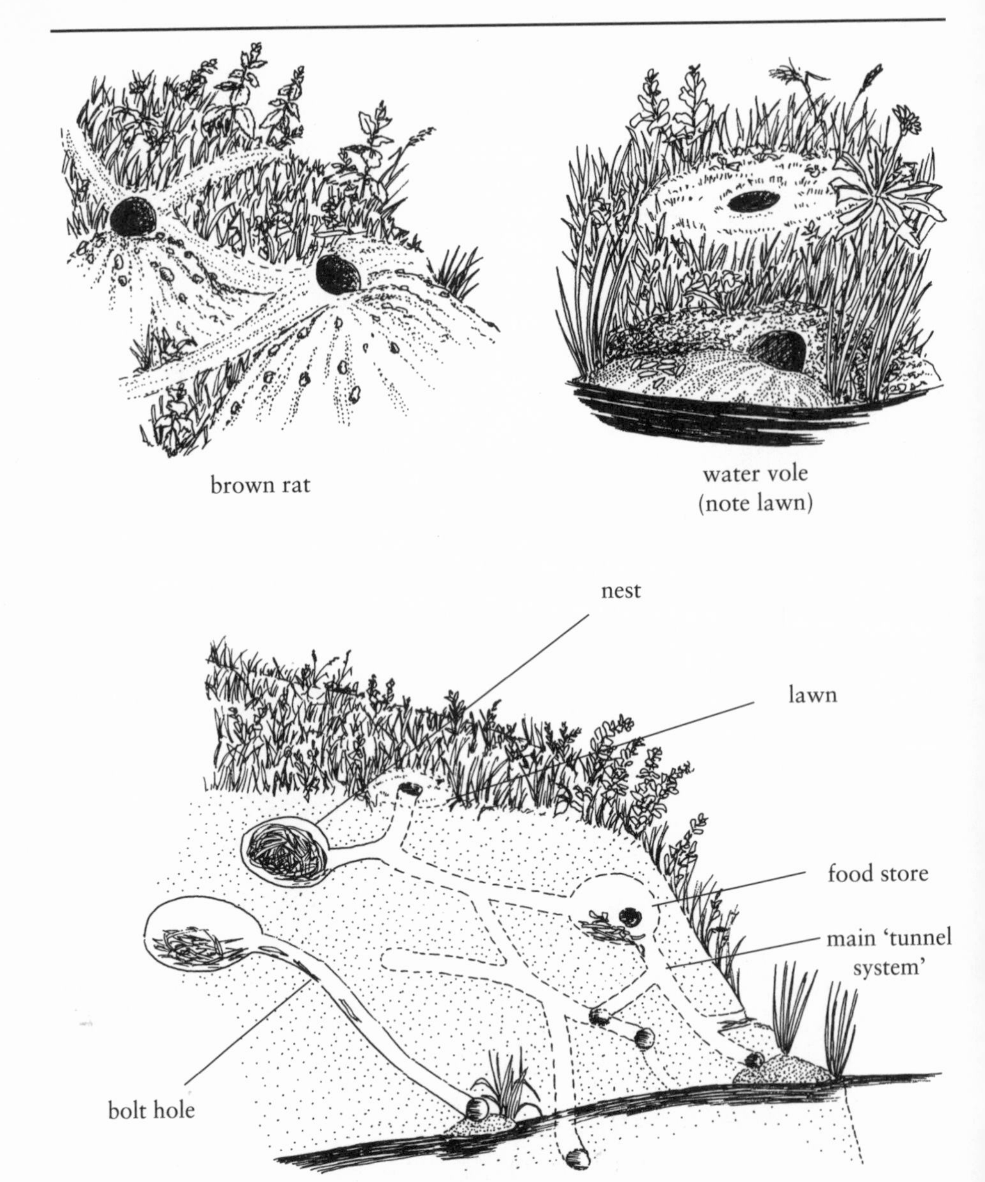


*Comparison of water vole and brown rat burrows. Note the fan-shaped spoil heap outside rat holes and the well trampled runs that link their holes. Water vole runs are less obvious, and closely grazed lawns may occur around holes on the bank. A cross-section through the bank reveals two types of burrow systems: (i) interconneting tunnels of a main system and (ii) single short bolt holes.*

ground, often woven into the bases of sedges, rushes or reed to form a ball. This large ball often resembles a moorhen's nest in size and composition but its inner core shows finely shredded material (the teeth marks at the end of each leaf blade, obviously not made by a beak!). When the water table is high, water voles may nest by hollowing out the centres of a greater tussock sedge or more rarely inside hollow willow stumps.

Both males and females take bedding underground to line nest chambers in the burrow system. Some nests are shared while others may only be used by single animals. Nurseries are particularly large and consist of a ball of finely shredded grasses or reed. Where these are underground, the chamber entrance may be plugged by the female with loose soil or grass in much the same way a rabbit does to protect the young from unwanted visitors.

**(v) Feeding stations** Food items are often brought by water voles to favoured feeding stations along their pathways or at haul-out platforms along the water's edge (as well as in burrow entrances, on the floor of the tunnels

*Where water table is high vole nests may be found woven into the bases of rushes, sedges or reed. Most nests however occur underground.*

*Chopped vegetation of around 10cm lengths can be found at favoured feeding platforms along the water's edge. By matching the pieces with the local plants growing on the bank the water vole diet can be determined.*

and in special undergound chambers). These show feeding remains as neat piles of chewed lengths of vegetation. The sections are typically about 8-10cm long showing the marks of the two large incisors and are quite good field signs of presence of water voles. The voles cut the stems to size for ease of handling and they will return to feed on them regularly. By matching up the plant fragments with what can be found growing at the site, you will be able to build up a list of preferred food and even observe how it changes throughout the year.

A pile of white pith stripped from *Juncus* rushes is another useful indicator of water vole activity, particularly in upland areas of Britain. Field voles will also do this, so look for the burrows and latrines to determine exactly which of the two species was responsible.

**(vi) Droppings and latrines** Perhaps the most distinctive field sign made by the **water vole** is its droppings. Usually 8-12mm long and 4-5mm wide, they are cylindrical with blunt ends and symmetrical in shape (looking a bit like guinea pig droppings for those of you who have them as pets). The colour varies from black, purple, brown to green, depending on the food eaten and the water content. The texture is like putty when fresh; when dry, they may show green concentric rings of fine plant material if broken open. They are generally odourless or perhaps give a faint musky smell when fresh and don't taste very nice!

Although a few droppings may be found scattered along runways, most are usually deposited at discrete latrine sites near the nest, at range bounda-

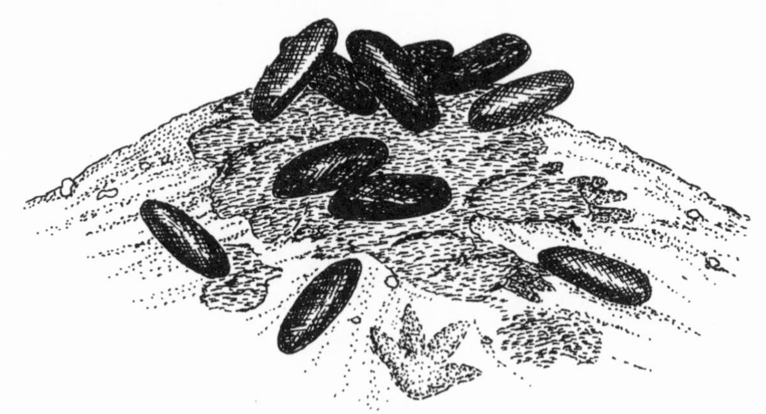

*Water vole droppings are typically left in an obvious pellet group at a latrine site.*

ries and at favoured platforms where they leave or enter the water. Latrine sites are typically established and maintained from February to November and play a role in the communication system during the breeding period. Scent from the lateral flank glands is deposited on the latrine when the vole drum marks with its hind feet, so that many latrines often show a flattened mass of old droppings, topped with fresh ones (see p. 34).

The number of latrines found and counted along a given length of waterway will give you an indication as to how many water voles you may have there (approximately six latrines will be maintained by one breeding female).

**Rat** droppings are rarely deposited in neat riverside latrines but are scattered along the runs, on ledges under bridges or on top of objects or riverbank features close to the burrows. Individual droppings are larger and fatter than those of the water vole at about 15-20mm long. They are generally blackish or brown, may be pointed at one end and have an unpleasant smell. (Always take care when examining any dropping because of the risk of disease, especially Leptospirosis or Weils Disease).

**Mink** droppings (known as 'scats') are usually 5-8cm long, cylindrical with tapered ends of less than 1cm in diameter. They may consist of fur, feather, bones and fish scales (occasionally crayfish shell or egg shell). Typically they are dark green or brown and covered in an unpleasant smelling secretion which once smelled is not forgetton. The scats are frequently encountered on prominences such as under bridges, on rocks and tree roots or around denning sites such as overhanging tree roots, log

water vole

brown rat

*Comparison of relative size and shape of water vole and brown rat droppings. (Life size)*

jams, hollow trees and rock screes.

The most distinctive field signs of the **otter** are its droppings (known as 'spraints'). These can vary in size from 2-7cm in length, about 1cm wide and cylindrical in shape. Black and tarry in structure and consisting almost entirely of fish bones and scales, they are smeared onto a prominent feature such as a rock, ledge under a bridge, a leaning tree trunk or among the exposed roots of riverbank trees. They are best told from mink scats by their very distinctive sweet fragrant smell, reminiscent of jasmine tea or laurel flowers, new mown hay with a hint of fish oil!

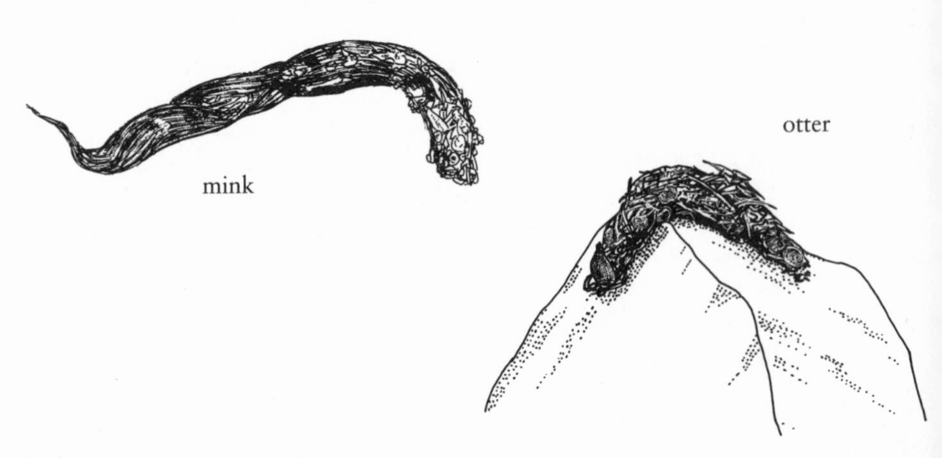


*Comparison of mink scat and otter spraint. Typically otter spraint is black, tarry and smeared onto a rock or other prominence. Mink scats are more cylindrical and contain more fur or feather.(Life size)*

# What do they eat?

*'Hold hard a minute, then !' said the rat.*

*He looped the painter through the ring in his landing stage, climbed up to his hole above, and after a short interval reappeared staggering under a fat, wicker luncheon-basket.*

*'What's inside it ?' asked the Mole, wriggling with curiosity.*

*'There's cold chicken inside it,' replied the rat briefly; 'coldtonguecoldhamcoldbeefpickledgherkinssaladfrenchrollscresssandwichespottedmeatgingerbeerlemonadesodawater——'*

*'O stop, stop,' cried the Mole in ecstasies: 'This is too much !'*

*'Do you really think so ?' inquired the rat seriously.*

Although Ratty may have favoured a traditional picnic for his lunch, water voles are really herbivores, feeding on the lush vegetation found along the riverbank. They consume the equivalent of up to 80% of their body weight daily. One published report I read, described a captive vole as daily requiring a two-gallon bucket full of juicy vegetation taken from a local brook. This vast quantity of food weighed two kilograms and was taken by a male weighing 285g: approximately seven times his own weight. I suspect what was really happening was that the vole took the plant material into underground stores.

From the remains found at feeding stations, it is possible to match up the various sections of plant stem with what is growing on site. Using this method the nationwide survey of 1989-90 (i.e. me) was able to identify 227 species of plant eaten by the water vole in Britain (and that was just from those bits of vegetation that I could readily find and identify!). This surprisingly large list has been compiled from the diverse number of habitats sampled and the wide geographic range for the species in Britain. At the site level there was a strong seasonal variation in the diet reflecting food availability and vole activity. Grasses, rushes and sedges are most frequently eaten and represent the bulk of the diet but in terms of the number of species on the list they form only a third of the total.

Pete Woodall, in his study of the distribution of water voles along parts of the River Thames during the 1970s, found that the best populations occurred where the fertile soils of the riverbank allowed dense stands of branched burr-reed to grow – a favourite food. Luxuriant growth of reeds, sedges, willowherb, water cress and other species offer food and shelter to

*Dense vegetation provides food and cover for water voles.*

water voles during the summer months.

However there is a change in diet over the winter period with roots and bark of trees and shrubs forming a major part of the menu, together with rhizomes, bulbs and roots of herbaceous species. Roots of various plants are eaten as the vole is tunnelling and the whole plant may be tugged down into the tunnel without the vole coming to the surface. At other times of the year aerial parts of the plants are the major components and in spring, leaf and flower buds may be taken. At some sites water voles were observed climbing into the branches of low growing trees and shrubs to a height of 2.5 metres and consuming the young leaves of hawthorn, buds of elder and leaves of willows. In the autumn fruits are also included in the diet (23 species were found to be eaten in the national survey). These are mainly from tree or shrub species which drop their fruits in this period, such as hedgerow crab apples. Apples are a particular favourite and in a few places water voles have been observed taking windfall fruits in gardens and orchards.

I once watched one cheeky individual repeatedly collecting apples from a riverside garden and then swimming across to its burrows on the other bank, pushing the apple in front of it.

Although they do take a few fruits, water voles normally prefer the coarser vegetative parts of plants rather than the higher energy fruits and seeds. This way direct competition with other rodents such as wood mice

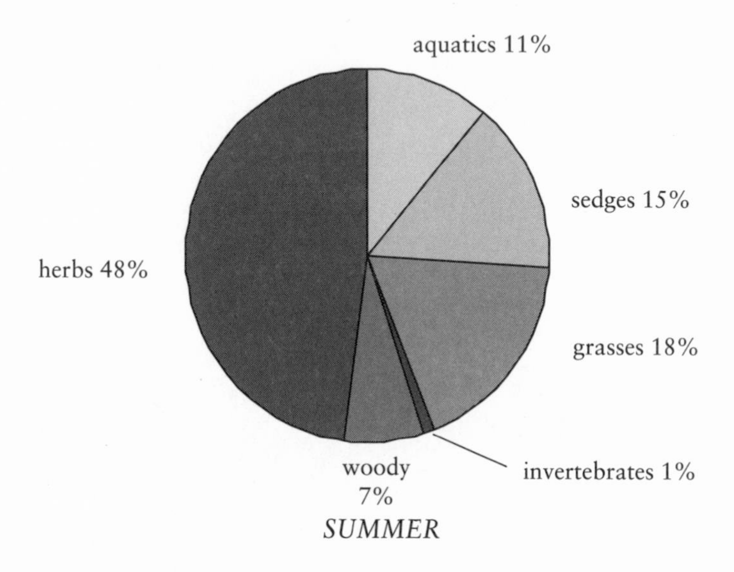


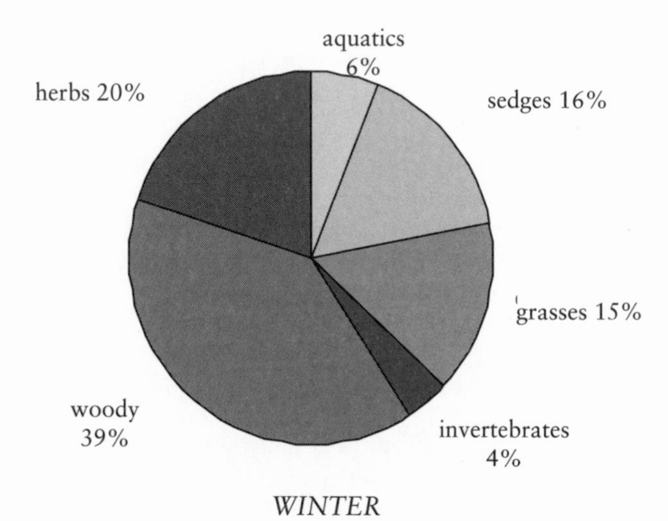


*Seasonal variation in the species composition of the water vole diet.*

*Water voles may climb to feed on young leaf shoots as they emerge.*

and bank voles is avoided.

Water voles may carry food back to the burrow system or a favoured feeding platform where they eat sitting on their haunches in a characteristic pose, holding the food item in their fore paws, feeding it into their mouths, like big sticks of celery. On some waterways the flowers of water crowfoot are readily consumed, the pollen providing a rich source of protein.

One spring evening I spotted a water vole on a small brook purposefully swimming out to collect fallen pussy-willow flowers that were floating by. With binoculars I could see that it was a heavily pregnant female and that she was only nibbling off the flower anthers. Finding a source of protein, such as pollen, may be essential for the female during the time of pregnancy. Some females will devour water snails and even crayfish at this time.

During lactation, the mother has to consume up to twice her own body weight in food each day.

In the winter months rhizomes, bulbs, cut stems and roots of herbaceous species may be collected into stores in their underground tunnels. One burrow system was found to have a hundredweight of potato tubers in its stores (a prudent householder getting ready for Christmas).

At a small number of sites water voles have been observed taking various species of freshwater mollusc and crayfish, apparently more often in winter

*Willows provide protein-rich pollen in spring and succulent bark over the winter (note teeth marks).*

(though perhaps remains were more easily found at this season).

Despite these notable exceptions the diet is almost completely vegetarian. From the list of species taken it can be seen that it is as wide as that found growing on site. It is possible that the presence of a wide range of species is not essential, though this does provide a 'guaranteed' succession of foods through the seasons. It is clear, however, that long and luxuriant growth is necessary and the water vole colony does best where bankside, emergent and water plant cover is high. It avoids places where overhanging trees and shrubs reduce the density and the number of species of the vegetation beneath it.

Relatively high vegetation cover allows voles to remain hidden while feeding, particularly from predators that hunt by sight, such as birds of prey.

*While feeding, the water vole assumes a characteristic pose, sitting back on its haunches.*

# Social life and behaviour

*'The bank is so crowded nowadays that many people are moving away altogether; O no, it isn't what it used to be, at all.'*

Water voles are not evenly distributed along a watercourse but show distinct habitat preferences and differences associated with overall population density and season.

**The colonial vole**

Water voles occupy a stretch of waterway in a social system that is loosely termed colonial. That is, they pack in side by side but defend exclusive territories within which the adult females each have their own burrow systems and breeding nests. Normally the territories arrange themselves along a waterway like a strung out line of sausages. The increasing day length of spring stimulates the onset of reproduction and the breeding season lasts from March to September.

Many of the individuals within a colony may be related but there is emigration and immigration between different colonies which may be

*Two water voles meet on a riverbank with much shrill chirping.*

separated by several kilometres of unoccupied waterway. On average a water vole colony may occupy between one and two kilometres of river length. During the winter the area used by the colony shrinks down, territorial behaviour stops and the voles peacefully share nests together. Over this period the voles' activity above ground is reduced and the animals show communal nesting between adult females and her female offspring and unrelated males.

**Defending territories**

When water voles that are defending territories meet they show particular antagonistic behaviour. Between two female voles, this begins with teeth chattering, flank scratching and tail beating or headlong chirping. If they are related, the loser backs down without physical contact; however, non-related females will fight overtly by boxing or wrestling (forelimbs about the muzzle and neck of the opponent) accompanied by much squealing. The interaction terminates when the loser breaks off and runs away, often with a bleeding nose or bitten ear. The frequent interactions between females determine social status and all low ranking females and juveniles are excluded from holding territories within that population. This encourages dispersal away from the mothers' territory for all except dominant daughters which may settle within or adjacent to her range.

Males on the other hand tend to be much less aggressive.

Home ranges extend in a narrow band along the banks of the waterway

*The challenger.*

*The flank gland is honey-combed in appearance and can be seen by parting the fur.*

and are often contiguous. The home range of males is about twice that for females and often overlaps those of one or more females. Females have only one mate, usually for the whole breeding season, but individual males exhibit polygamy (that is they may have more than one mate) and share their time between the different territories accordingly.

When the population density is low, males occupy ranges of up to 300 metres and breeding females up to 150 metres length, but when the population density is high the mean length of range is much smaller at

*The territory is challenged.*

*Territory is marked by stroking the hind foot across the gland and stamping on the ground.*

around 100 metres for males and 50 metres for females. Surprisingly, under conditions of high total population densities, breeding females may overlap in their ranges and tend to be less aggressive to one another.

The ranges are marked by discrete latrine sites and at the boundaries of the female's territory large latrines can be found where their mate scent marks. Christina Leuze demonstrated experimentally that if she removed the adult males from their mate's territory but she artificially maintained the end latrines with his droppings, the male scent stopped the females

*Boxing and then rolling into a biting ball.*

from extending their ranges beyond that boundary point. However, if she removed the end latrine as well the resident females attempted to extend their territories, resulting in aggressive encounters between neighbours. These latrines are composed of flattened masses of droppings topped up with fresh ones, since the water vole scent-marks at the time of defecation by stroking the hind feet across the lateral scent glands on its flanks and then drumming them on the latrine.

These glands are roughly oval in shape, up to 18mm x 12mm in size and look like a rugby ball lying on its side. Their surface is slightly raised above that of the surrounding skin and has a honey-combed or irregularly wrinkled surface, almost hairless (they can be seen very easily when handling a water vole, by blowing on the flank to part the hair) and dark brown in colour. A thick waxy deposit is secreted from the glands and builds up on the surrounding hair; it has a distinctive scent that the voles use to communicate with one another.

It is generally believed that the flank glands of water voles produce socially important odours. These are used to mark latrine sites and other features (such as a prominent tree root exposed at the water's edge) within each animal's territory. It is very likely that the chemical composition of the odours inform other voles about who's territory they have strayed into, whether they are male or female, whether the female is ready to mate and possibly information on how healthy the animal is and what its social rank is. They therefore play a very important role in water vole society.

*The victor.*

They may even tell of the richness of water vole diet, indicating how good that particular territory is and what habitat it includes – and so whether it is worth fighting over!

**Breeding**

Barbara Blake observed the courtship behaviour of water voles in a captive colony and found that it was initiated by the male drumming with its hind feet on top of the nest box (Phil Collins 'in the air tonight' perhaps), scent marking behaviour that has also been observed prior to mating at boundary latrines in the wild. The male would also make soft audible chattering calls and ultrasounds that could be detected electronically (such as with a bat detector). This way he would attract the female and he would try to approach her for mating. This would result in several small skirmishes, the female rebuffing the male with snarling and striking out with its forepaws, running the claws up and down to scratch the male before he would be successful. Mating may take place on the land or in the water and often the male deliberately knocks the female into the water, chasing her to tire her out. At last, exhausted and catching breath at the water's edge, the male mounts the female unchallenged.

After mating the male and female may share the same nest until she is ready to give birth, whereapon she chases him out to a separate nest chamber in the same burrow system. Once the young are born, the male plays little part in rearing them but does appear to recognise his progeny and may even nuzzle the little ones as they run about the territory. If the nursery

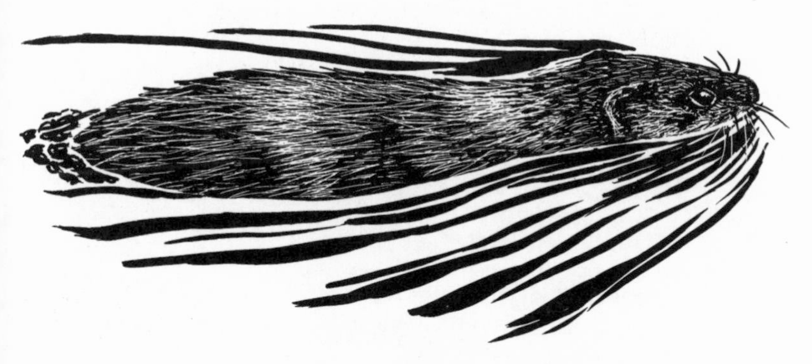

*The challenger is vanquished.*

nest is disturbed both the male and female will rescue the young and carry them off to a safer refuge in an alternative nest.

Litters may consist of 2 to 6 young (exceptionally 9 or 10), 5 being the more usual at the beginning of the year, the numbers becoming less as autumn approaches.

The young are born blind, hairless and toothless, with small ears, and are generally helpless. They develop rapidly on the rich milk from the mother as she suckles them lying across the nest. After five days the young (known as pups) are well furred, covered in a downy golden red fur. Their eyes open at about 7 days and by 10 days they have thick fur and the full appearance of the water vole (although they resemble field voles in size). They usually stay in the nest until around 15 days old, following closely behind the mother, even taking to the water to bob along behind like fluffy corks. They are fully weaned at 28 days.

**Vocalisations**

Marion Browne kept a few water voles at her home in Wiltshire in the early 1980s and made notes on the sounds they made. She described a distinctive chirping call when the animals were agitated or annoyed (such as when she was trying to round them up for clearing out their pens). The call had a definite rhythm, all on one note, that was reminiscent of the chirping of a house sparrow. Six notes repeated, 'one and two, three, four, five – pause – one and two, three, four, five', the 'and' being a short chirp.

Other calls have been shown to be a prelude to mating or fighting and as contact calls between mother and her young. Some of the latter are above human hearing but can be heard on a bat detector set at about 35KHz.

**Activity patterns**

Observation of water vole activity has shown that they follow a roughly 4-hourly rhythm, being seen foraging above ground for about one hour at three-hour intervals, although they can be more active at dawn or dusk. Radio-tracking studies have found that although above ground activity may follow a cycle, they do not simply spend the rest of the time asleep in an underground nest. They are continually moving about below ground, perhaps repairing or extending tunnels, creating nests or food stores or eating from those stores. Females nursing unweaned young in the nest will make frequent return visits to them, limiting excursions to sometimes just reaching out of the burrows to graze the nearest vegetation into a short cropped lawn. They will dart back underground at the first sign of danger. An observer sees a little round nose poking out of a hole up on the bank, scenting the air, then the vole quickly reaches out, grabs a mouthful of food and darts back down again. The whole process will be repeated a few seconds later.

**Dispersal**

Dispersal away from the place of birth allows for new areas to be colonised or new blood to arrive in an existing colony by way of immigration. By their very nature these long distance movements are difficult to study and monitor. However, the fact that it does happen regularly has been shown in colonies that are being studied by intensive trapping and marking of individuals; unmarked new adults have appeared in such colonies. Similarly, already marked individuals may be recaptured elsewhere, as demonstrated by Michael Stoddart in his detailed study of the water vole population at the Sands of Forvie Nature Reserve, near Aberdeen, Scotland. His study described short distance movements as either 'sallies' outside of the normal home range that lasted for only a day or so before the animal's return to its original site, or permanent movements to a different territory within the colony. Long distance movements were recorded away from the colony to a totally new area (see illustration). These movements were observed among juvenile males and females, but also some adult females. Adult males were found to remain faithful to their individual range over the whole of their lifespan. Michael reports that one female was first caught as a breeding adult in June on a small 'Y'-shaped stream in the south-eastern part of the study site. She was recorded as having one pregnancy there until the middle of July, after which she disappeared. By chance, she was recaptured 9 weeks later nearly 3km to the north of the stream, occupying a new territory on a different stream and rearing another litter of pups there. To get to this new site she must have crossed an area of open rough grouse moor, perhaps by moving between the damp swampy patches on her journey.

**En-suite latrines**

Radio-tracking studies have demonstrated that voles maintain a territory

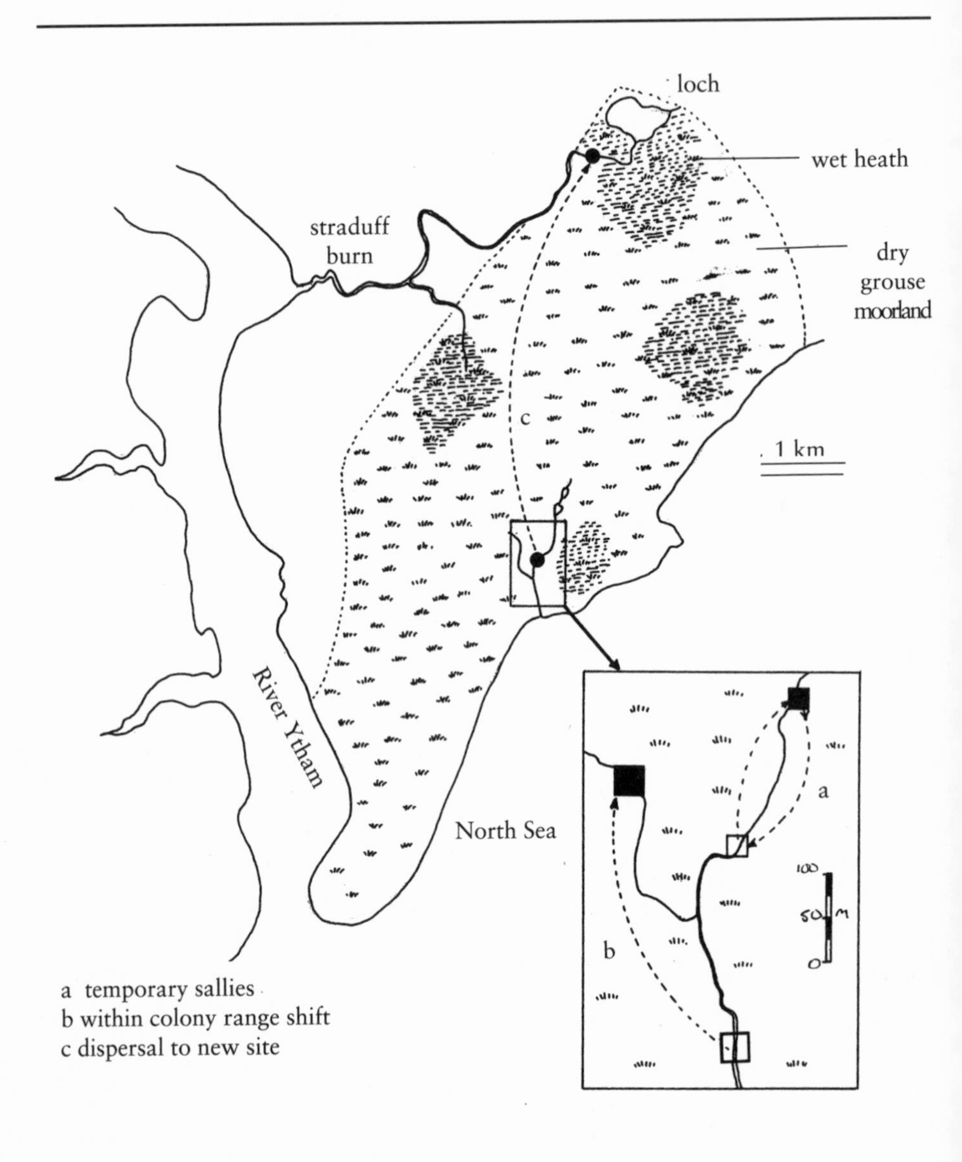


*Movement of water voles within and between colonies on the Sands of Forvie Nature Reserve, near Aberdeen (after Michael Stoddart 1970).*

system and that they use latrines as an important part of that system. These latrines are daily visited by the water vole and can be readily recognised by their large size. There is a direct correlation between the number of latrines and the number of animals at breeding sites. One study by Gordon Woodroffe, looking at the water vole density at the breeding sites on rivers in the North York Moors National Park, found a density of 10 animals per 300 metres of river bank, a much lower density compared to other trapping studies carried out in other habitats. For instance, Christina Leuze located 50 animals over only 825 metres at Bure Marshes, Norfolk, and John Singleton trapped between 19-40 animals over 800 metres, varying with season, at a lowland dyke system in West Lancashire. Thus, the numbers of water voles trapped within a 100 metre stretch of suitable riparian habitat ranged from 2.37 individuals (West Lancashire) through 3.33 (North Yorks Moors) to 6.06 (Bure Marshes) , depending on the quality of habitat and time of year.

In continental Europe, however, the water vole density may far exceed the figures given here. J.Pelikan & V. Holisova determined the presence of 6-9 animals per 100 metres at a low lying brook in Russia.

# A plague of voles

*The burrowing form of water vole is known to occur at densities of between 200-500 per hectare, causing damage to orchards and bulb fields in France, Switzerland, West Germany and the Netherlands. In the Kocansko Valley of Yugoslavia an 'explosion' of these animals among rice fields reached a density of 1,000 individuals per hectare in 1987 when they were controlled by poisoned baits. Plagues of water voles have been extremely rare in this country and there has only been one well documented case. This occurred one hundred years ago, back in 1896 on Read's Island, in the Humber estuary near Hull. It was first thought that the island was being over run with brown rats, but subsequent investigation proved to be water rats. At that time Read's Island was described as 600 acres (242 hectares) of clover and grass ley, capable of supporting 3,000 sheep and cattle. Flat as a billiard table, the island was surrounded in its five miles of perimeter by three metre high earth banks to keep the sea out.*

*A cutting from the* Eastern Morning News *graphically described the outbreak: 'The Humber Conservancy*

*Commissioners have not yet solved the difficult problem of how to rid Read's Island of the plague of water rats that now infest it... It is burrowed from end to end, and so densely populated is this habitat of the rodent that it is said that it is almost impossible to put a foot down without standing on a rat-hole. The entire island is as brown and rough as a ploughed field and there now exists scarce sufficient pasture to feed one rabbit.' On flooding the island at spring tide by creating cuttings in the earth bank defences and fitting sluices to retain the tide, the Commissioners observed that 'as the water advanced the rats fled from their holes in tens and hundreds of thousands, and made for the banks which remained high and dry. Many were doubtless drowned by the flood but, by being expert swimmers the impression made on this great army was practically nil.' Next a shooting party was invited for a day's sport and although 'hundreds were killed' it made little impression. Starvation eventually caused the population to crash, with many voles swimming across the Humber estuary to the mainland. Read's Island was devoid of rabbits, moles and brown rats at the time of the 'plague' and the water voles' success may have been in part due to the absence of these animals, the water voles living in a way reminiscent of all three.*

*Read's Island never saw another similar outbreak, although some water voles could always be found on the island (until the 1980s).*

*Juvenile water voles appear very fluffy and not as sleek as adults.*

# Scientific studies

Capture of water voles may be necessary for detailed scientific study or for conservation reasons such as removal from a site threatened with destruction. In both instances, check whether a licence is needed from the relevant Statutory Nature Conservation Organisation (i.e. English Nature, Scottish Natural Heritage and the Countryside Council for Wales).

Water voles are relatively easy to capture, especially during the breeding season when they are frequently patrolling their home range. The key to a successful trapping exercise is where you put the traps and what you use as bait.

Traps available for this work are of the cage design with a treadle mechanism that releases the door held open against a spring. The dimensions are approximately 10x10x30cm for the tunnel, with a nest box attached.

**How to catch a vole**

Between April and October, latrines are regularly visited by resident water voles and if you place a trap immediately beside a latrine success is almost guaranteed. The trap should be secured so that it doesn't tumble into the water when a vole is caught and should be placed so that it is not at risk from submersion by rising water levels. A small channel dug into the bank at right angles to the watercourse may provide a long term capture site, but if you intend to dig the trap into the bank permission must be sought from the landowner. Traps should be baited with sliced apple and carrot (about 200g) and bedding of hay provided if the trap has a nest box.

The traps should be checked at least twice a day and preferably every four hours.

**Handling the captures**

Water vole are easily stressed and they have a very strong bite, so they will need to be handled with care and with leather gloves for your own protection.

If the voles are being caught for safe removal to another site (such as in a translocation programme) then they could be left in the trap or transferred to a transportation box (a cardboard box type pet carrier as available from pet shops) keeping each animal separate to reduce risk of injury through fighting.

If the voles are being caught as part of a scientific study, then good scientific animal care in handling should be adhered to. The water vole is transferred from the trap to a soft net bag (such as the poly-mesh material as used in keepnets and landing nets by anglers) in which it can be weighed and examined. The sexing of breeding adults is obvious, swollen mammary glands in the female confirm breeding and lactaction, but in non-breeding or juvenile animals when the organs have not descended from the body cavity, then the distance between the anal-genital openings needs measuring. For males the distance exceeds 10mm (gently everting the penis with the finger tips confirms the sex), in the female the distance between the anus and the vaginal opening is less than 10mm.

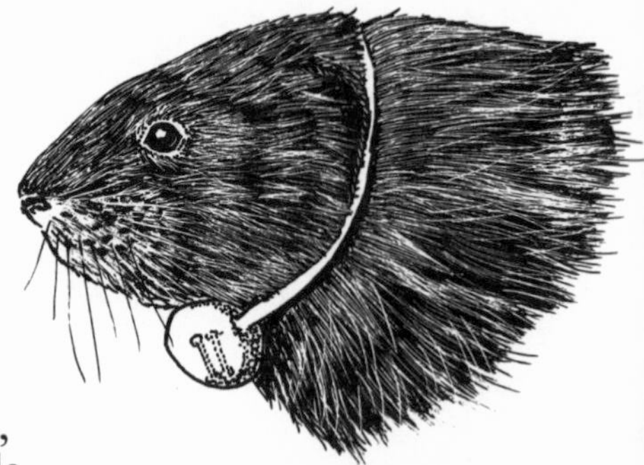

*A water vole sporting its new water proof radio-collar!*

For the more detailed examination and for the safe attachment of a radio-collar anaesthesia is recommended (such as inducing anaesthesia in a flow chamber with a volatile anaesthetic of a wide safety margin for rodents). Under the Animals and Scientific Procedures Act 1986 a Home Office licence would also be required.

**Radio-tracking**

Radio-collars consist of a small battery powered transmitter at a set frequency in the 173-174MHz waveband (designated for use with British wildlife). Each radio-collar has a specific range of 2-3KHz allowing a large number to be used in close proximity without interference from one another. The button cell battery and transmitter are sealed in resin and then coated in a plastic to make the unit waterproof and a 10cm long whip antenna increases the range over which the signal can be sent. The button cell battery has a life span of between 20-40 days. The unit is then attached to a cable-tie as a fastening to go around the vole neck. The total weight of these small collars is around 3-3.5g and so well within 5% of the water vole body weight, and therefore not likely to be burdensome to the vole. Collared animals soon settle down and go about their daily business undeterred, allowing the scientist to record what is going on without getting so close that the animal is disturbed.

A directional aerial and receiver allows the signal to be picked up over a

distance of up to 200 metres line of sight.

The technique of radio-telemetry is most useful in determining the activity pattern and movements of the water vole as well as the limits of home range. Pending legal protection, a research licence must be sought from the relevent Statutory Nature Conservation Organisation (see addresses at back) if you intend to trap or radio-track water voles.

**Dietary studies**

Studies on the diet of the water vole can be accomplished in a number of ways:

- by direct observation of the voles chewing plant material
- by collection of remains found at feeding stations (these can then be matched up with the plants found growing in the proximity)
- by the microscopic examination of droppings. The latter is the most difficult and time consuming, but does allow the vole diet to be studied where feeding remains have not been found and where voles are rarely seen. The droppings can be dried and then teased apart on a microscope slide or dissolved in water to make a soup droplet that is placed on a slide under a coverslip. The tiny fragments of plant material are then

examined for the shape and size of epidermal cells and their associated stomata. These are the specialised cells on the surface of the leaf through which the plant transpires. Under the microscope different leaf surfaces can be identified by the shape, number and arrangement of these cells (see drawing).

The seasonal succession of the plants in the water vole diet may be important to the long term survival of the water voles at a site. Food shortages may occur, particularly during the winter period when new plant growth is scarce.

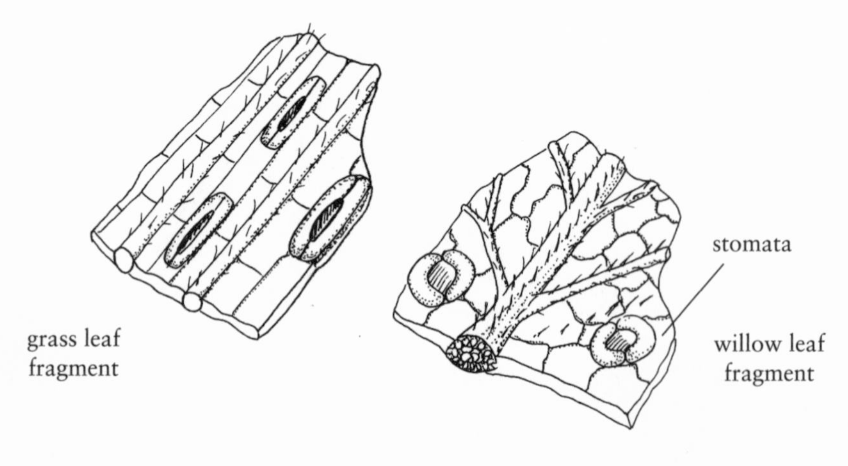


*If you tease vole droppings apart under the microscope, you can recognise plant fragments by the size and shape of the surface cells and stomata. Grasses and other related species have regular shaped cells running in parallel lines, while broad-leaved plants have typically irregular cells and circular stomata.*

# The water vole calendar

*'When the floods are on in February, and my cellars are brimming with drink that's no good to me, and the brown water runs by my best bedroom window; or again when it all drops away and shows patches of mud that smells like plum-cake, and the rushes and weed clog the channels, and I can potter about dry-shod over most of the bed of it...'*

Winter or summer, rain or sunshine, Ratty has learnt to live with the vagaries of the British weather and changing seasons.

**January-February:** Very few obvious signs of activity can be found as the population density is low and the majority of time is spent below ground feeding on stored food in their burrows or on roots and rhizomes. Where vegetation cover remains dense, such as in *Phragmites* beds, extensive runs can be found in the litter but nesting is underground, often communally.

**March-April:** Females determine onset of breeding. Nesting females space apart into non-overlapping territories. Latrines established and regularly scent marked. Adult males associate more with females and many overlap with one or two mates. Immigration makes up the complement of breeding females in the population.

**April-May:** Peak in birth rate. Gestation period lasts 22 days and litter size ranges from 2-6, averaging 5. Sex-ratio of the litter is 1:1. Within a few days of a female giving birth to a litter she may be ready to mate again.

**June:** Peak in newly weaned animals, when the juveniles show the first excursions from the nest at 25 days old. The young animals may be seen swimming very buoyantly across the water surface, bobbing about like corks. They may be less than half the body length of the adult and weigh around 40g.

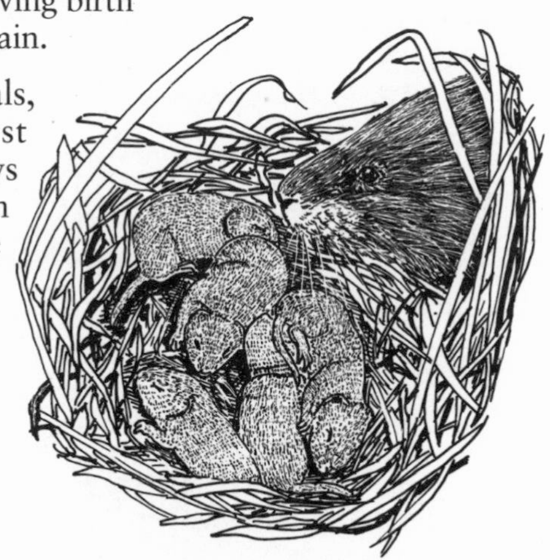

*Five or six babies are born with their eyes shut and covered in downy hair.*

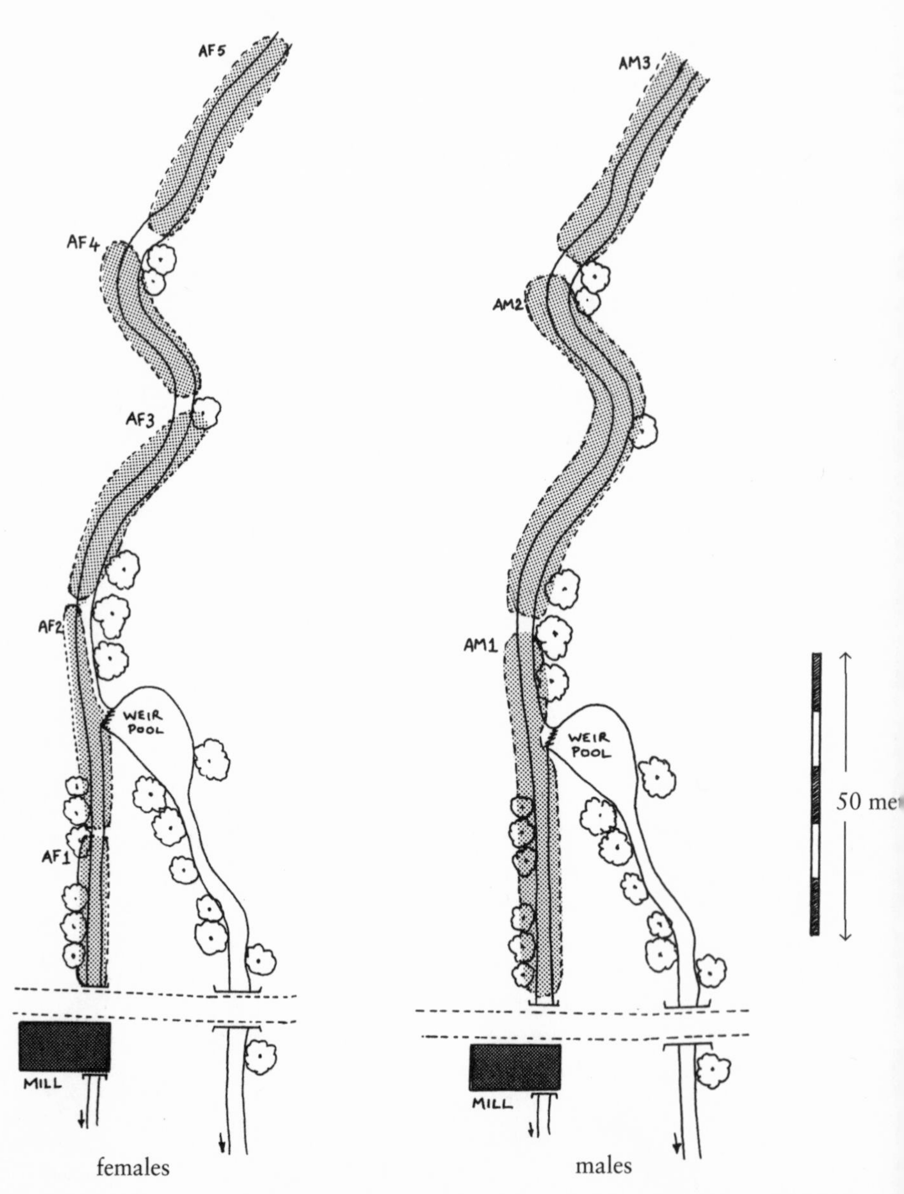


*Territories separate out along a water course like a string of sausages. On the River Windrush in Oxford the males overlapped the ranges of two females. Adult females (AF) occupy a range of around 40m (both banks on this narrow river), while adult males (AM) hold a range of 80m (author's data).*

**July-September:** Second and third litters produced (exceptionally up to five litters a year). Dispersal of juvenile males and low ranking females. Dominant daughters may settle inside mother's territory and may even displace pregnant female following a territorial dispute. Nests can be found above ground, often woven into the bases of sedges.

**September:** Population size at an overall peak. Females born early in the year may become sexually mature and breed but the majority will not breed until the following year.

**October – November:** Reduction in numbers by dispersal. Preparation for winter by laying down underground food stores such as harvested hay. Sexual activity ceases and territorial disputes become less frequent.

**December:** Territorial system now broken down as home ranges contract. Although no evidence of hibernation, some individuals enter short periods of torpor in subterranean nests followed by bursts of above ground activity. Low survival rate among juvenile animals whose overwintering weight averages 170g. The voles now gather together to share nests.

# Local names

*Not only has the water vole had many scientific names, it has also figured in folklore and enjoyed many local names, including water rat, water ratten, water mole, craber (from the Breton* raton-crabier*), waterdog, earth hound and water campagnol.*

*It is sometimes called* Llygoden y dwfr *and* Lamhalan *in Gaelic.*

*One old folk story from Aberdeenshire and Banffshire recounts that the water vole in its black form went under the name of earth-hound, a mythical creature supposed to frequent graveyards and devour the dead! Although an unlikely story, its origin lies in the fact that some water voles have been caught in graveyards close to rivers; molecatchers have found them in traps set in mole runs below ground.*

# Water vole surveys

In 1989-90, The Vincent Wildlife Trust carried out the first ever systematic National Survey for the water vole, visiting 2,970 different localities in England, Scotland and Wales. The aims were fourfold:

(i) To determine the distribution and relative status of the water vole in the United Kingdom by systematic search. (All previous maps relied on records being sent in from interested people and often had poor coverage of the country, the distributions often mapping where an active mammal recorder lived rather than giving a true picture of the animal's actual whereabouts).

(ii) To relate presence/absence of the species with habitat features and presence of feral mink.

(iii) To provide a series of referenced 600m sites to act as a baseline from which to measure any future stability, increase or decline.

(iv) To examine a series of sites with reported previous dated presence of water voles this century so that there would be some immediate indications of whether a decline had occurred, whether this was long or short term and the area of greatest effect.

The 1989-90 Water Vole Survey was designed to have two approaches to answering the question of how common and widespread they were.

The first approach was what is called a baseline survey: visits were made to a series of about 2,000 pre-selected sites which then formed the country-wide baseline from which future repeat surveys could measure any declines, increases or stability in the population as well as any changes in distribution.

The second approach was an historical survey: visits were made to a series of about 1,000 sites known to have had water voles at various times in the past so that these could be checked for present habitation. This would then give an immediate indication of population change and its time scale without waiting for the second survey seven to ten years into the future.

**The baseline survey (a systematic search)**

Pre-selected series waterway sites spread over mainland Britain were

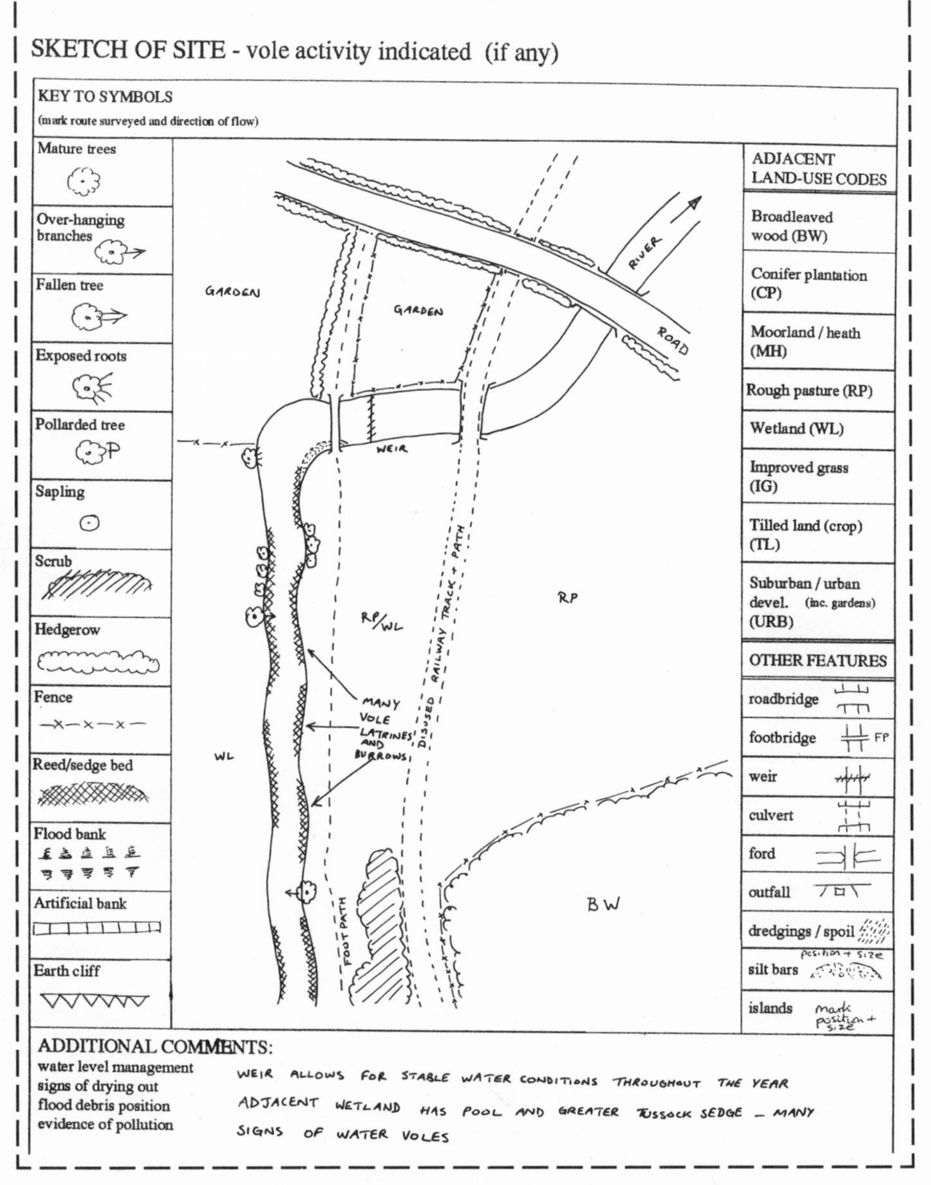

SKETCH OF SITE - vole activity indicated (if any)

KEY TO SYMBOLS
(mark route surveyed and direction of flow)

Mature trees
Over-hanging branches
Fallen tree
Exposed roots
Pollarded tree
Sapling
Scrub
Hedgerow
Fence
Reed/sedge bed
Flood bank
Artificial bank
Earth cliff

GARDEN
GARDEN
RIVER
ROAD
WEIR
RP/WL
RP
DISUSED RAILWAY TRACK + PATH
MANY VOLE LATRINES AND BURROWS
WL
FOOT PATH
BW

ADJACENT LAND-USE CODES
Broadleaved wood (BW)
Conifer plantation (CP)
Moorland / heath (MH)
Rough pasture (RP)
Wetland (WL)
Improved grass (IG)
Tilled land (crop) (TL)
Suburban / urban devel. (inc. gardens) (URB)

OTHER FEATURES
roadbridge
footbridge FP
weir
culvert
ford
outfall
dredgings / spoil
silt bars position + size
islands mark position + size

ADDITIONAL COMMENTS:
water level management
signs of drying out
flood debris position
evidence of pollution

WEIR ALLOWS FOR STABLE WATER CONDITIONS THROUGHOUT THE YEAR
ADJACENT WETLAND HAS POOL AND GREATER TUSSOCK SEDGE – MANY SIGNS OF WATER VOLES

*A completed survey form sketch map depicting the habitat and frequency of water vole signs (author's data).*

sampled, with equal effort being placed in each area. A grid of the same 18 10km squares (i.e. 02,05,09,20,24,27, etc.) in each 100km square of the National Grid was chosen. Within each of the selected 10km squares 5 sites were chosen from the Ordnance Survey map which gave a good representative sample of the waterways available. This gave a total of 1,926 different sites.

Each individual site had an accessible starting place, such as a bridge, for which a grid reference could be given. These sites would be a canal side, river, tributary, stream, ditch, dyke, gravel pit, lake, loch or reservoir, etc.

These were then visited and searched for a distance of 600 metres along the watercourse bank (one bank only). The field survey was usually divided into 300m upstream and 300m downstream of the access point or bridge. The signs of water voles were recorded on standard survey forms (see p. 49) at every site: such things as sightings, sounds entering water, latrines showing discrete piles of droppings, tunnel entrances (above and below water), cropped 'gardens' or 'lawns' around tunnel entrances, feeding stations of chopped vegetation, paths at water's edge, runs in the vegetation and footprints in the mud were noted.

The numbers of burrows and latrine sites were counted in the 600m field search to provide an indication of relative density of water voles at that site for comparison between sites and future surveys.

Food remains were also identified to determine the species being eaten.

For each site standardised habitat data were recorded by ticking relevant boxes on the survey form. Information was gained on type of waterway/ water body, water flow, width, depth, shore type, emergent and bank vegetation, dredging activity, drainage works and canalisation, bordering land-use/adjacent habitat and water use. Signs of waterside recreation (boating, angling, etc) and apparent level of disturbance was also noted. Each site was assigned a survey number and name, together with an accurate grid reference, date of visit, altitude, county and water authority region.

A sketch map was also drawn showing which bank was surveyed to facilitate repeat surveys in the future.

The presence/absence of mink signs at each site was also recorded on the data sheet, with the relative abundance of footprints and droppings (known as scats) located along the 600 metre distance being noted.

Thus, the survey doubled as a mink distribution survey and allowed for the relationship between water voles and mink to be examined (see p. 61).

Other wildlife such as certain water birds and otter signs were also recorded when encountered but they were not actively searched for to the same extent as mink.

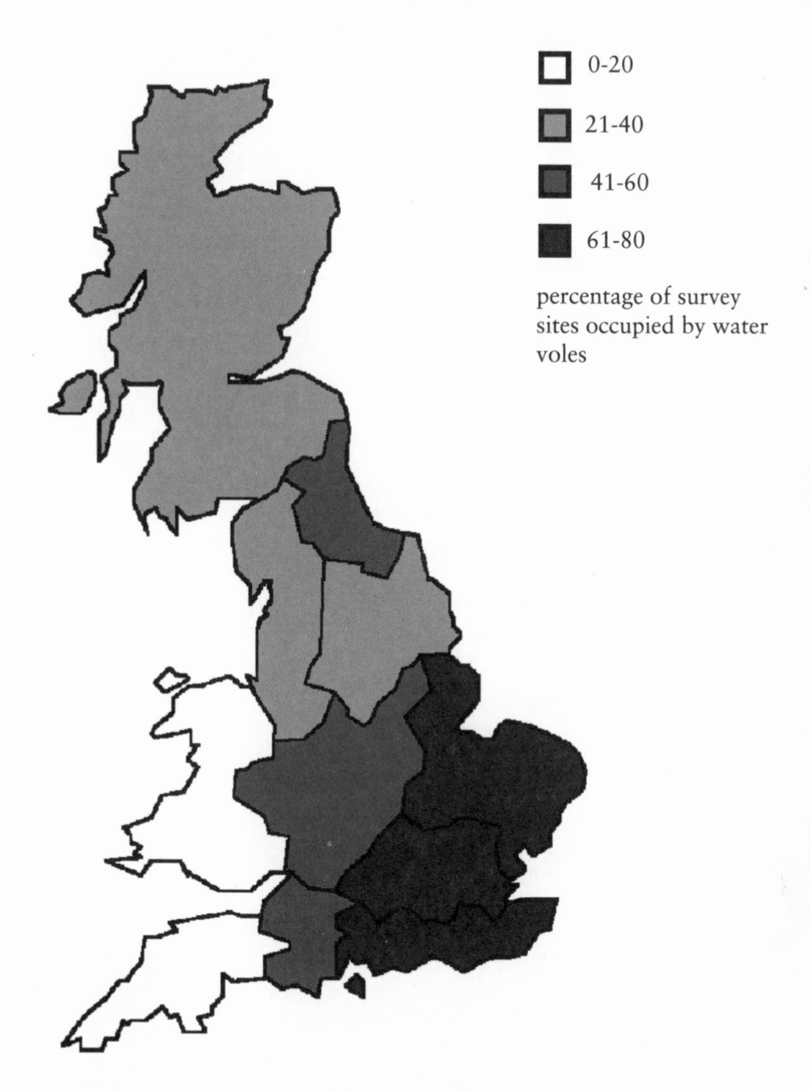


*Relative distribution of water voles in each Environment Agency region as determined by survey during 1989-90 (based on the Vincent Wildlife Trust report* The Water Vole in Britain 1989-90*).*

**The historical survey (a directed search)**
Old records and reports were searched for accurately referenced and dated sightings of water voles. These included Biological Recording Centre Mapping Scheme (at Monks Wood, Cambridgeshire), County Museum Records, British Trust for Ornithology Waterways Survey, Otter Surveys, Local Mammal Reports and information from local naturalists.

The records were divided into six time periods: sites occupied in 1900-1939, 1940-49, 1950-59, 1960-69, 1970-79 and 1980-89.

Where possible sites falling outside of the baseline 10km squares were chosen to improve the distribution information. Of these sites 1,044 were visited again during 1989-90 and surveyed as for baseline sites above, checking for the continued presence of water voles in each time period after an interval of 10, 20, 30, 40, 50 or up to 90 years.

Additional records and reports were sought through information requests in various magazines and journals (including County Wildlife Trusts, Local Natural History Societies, Mammal Society, *British Wildlife Magazine*, *Country Life Magazine*, *Women's Institute Magazine*, angling and boating magazines, national and local newspapers and Radio 4's Natural History Programme). These records largely complemented the data on distribution gathered by field survey work and helped provide comprehensive coverage of Britain. This made the water vole survey one of the most detailed mammal surveys ever carried out in Britain.

Of the sites visited, 47.7% were found to hold populations of water voles, but where water voles had been previously recorded before 1939 only 32.3% were still occupied by them in 1990. This survey confirmed the steady long-term decline of the water vole populations but also highlighted that the rate of decline appeared to be accelerating through the 1980s and 1990s (see p.58). The survey report predicts that if the rate of decline continues then by the year 2000 we would have seen a total loss of 94% of occupied sites this century (so only 6% remaining).

The 2,970 survey sites form a base of information against which future trends in the fortunes of the water vole can be measured. At the time of writing, all 2,970 sites were being re-visited over the two year period 1997-98 by The Vincent Wildlife Trust.

# Life on the road

*I still find it amazing that the 1989-90 survey was carried out just by myself, living and travelling around Britain in a camper van. Despite the daunting task of visiting nearly 3,000 sites, it had been anticipated that a single surveyor could apply constant effort throughout, allowing even coverage, particularly in areas supporting few or no water voles.*

*The work schedule meant that I started in January 1989 and finished in December 1990, working every month. The survey was designed so that I visited low-lying parts of southern England in the winter months where the weather was generally milder and voles could still be found active and that I visited upland and northern Britain in the summer months when conditions were best for*

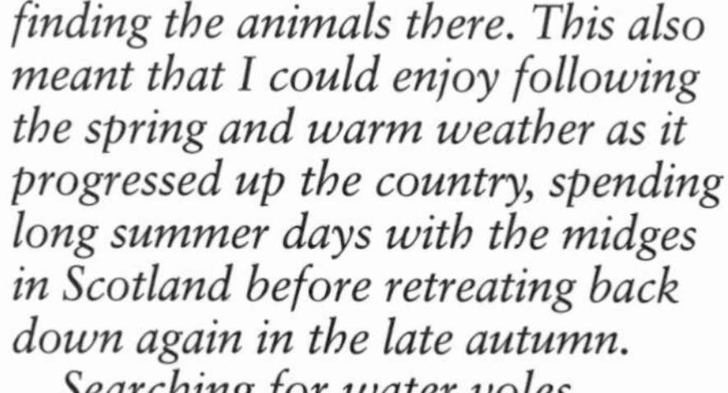

*finding the animals there. This also meant that I could enjoy following the spring and warm weather as it progressed up the country, spending long summer days with the midges in Scotland before retreating back down again in the late autumn.*

*Searching for water voles between Land's End and John O'Groats, I was able to find the species where they had been overlooked before, such as among the streams and mires of Rannoch Moor or the Flow Country of Caithness and Sutherland. Previous distribution maps failed to show water voles there because few mammal recorders had visited and noted them (and so no previous dots on maps).*

*The survey also found voles reaching new heights and altitude records! One small local population of water voles was found at the uppermost reaches of the Elan-Wye in Wales, to an altitude of 450m while another was located in good numbers at a high plateau site of the Cailness burn, 550m up Ben Lomond, in Scotland.*

*West Highland systems including Torridon, Ewe and Carron were found to support very good populations on small meandering tributaries. I found one population on the River Torridon of black and tan. They are probably still there!*

**Local surveys**
Local surveys can be more detailed than the overview of a national survey, because of their generally smaller scale, incorporating a greater frequency of sites along each watercourse or, better still, examining an entire waterway system metre by metre.

Catchment-based surveys provide a regional overview, which ignore the artificial constraints of political boundaries of each county but are determined by local geography and hydrology – more natural systems that have a direct bearing on riparian mammal species.

These are termed 'hydrometric' areas and may be surveyed in their entirety to give a complete and accurate picture of a species distribution and so demonstrate the integrity of the riparian habitat corridor. This will then identify gaps in the habitat and the level of fragmentation of the water vole population – and ultimately direct the focus of conservation effort.

The best time to conduct field work is late April through to early October, when water vole breeding territories are marked by latrines.

# The Oxford elite

*During 1996, I co-ordinated the first ever detailed survey of the waterways of Oxford City in order to see if any water voles remained. Some 75km of river, stream and canal wind their way through the city boundaries and the entire length was surveyed by local people in a City Council sponsored community project. The waterways were divided up into adjoining lengths, each 500m long and these were allocated to volunteers who had attended a training day on 'how to look for water voles'. Well over sixty volunteers took part and all the waterways were visited; five separate discrete vole colonies were identified. Each colony will now have more detailed habitat management work done to secure its future as part of the City Nature Conservation Strategy.*

*Why not work with your local Wildlife Trust, to get a local survey off the ground? You will be surprised at how many people are willing to help and the results could make a real difference by focussing conservation work to help the voles.*

**Canal surveys**

The national water vole survey of 1989-90 found water voles at 90% of the canal sites visited, demonstrating the high suitability of slow flowing canals with stable water depth, provided they had natural banks rather than metal sheeting or concrete sides. To find out more about canals, British Waterways carried out a survey in 1996, among the thousands of narrowboat users, asking them to record water voles whenever they saw them, on its Waterway Wildlife Spotter's Card as they travelled these inland waterways. Although canals are artificial waterways, significant colonies of water voles were located on the Kennet and Avon, Oxford and Staffordshire and Worcester canals.

These results will be used to produce detailed mapping of water vole colonies along canals to enable British Waterways to take these habitats into account when planning maintenance and repairs.

While cruising on the Oxford canal in a pea-green narrowboat called *The Owl and the Pussycat*, a friend of mine would eagerly count the water voles along the long sections between flights of locks. Where both the towpath and opposite banks were fringed with sedge or burr-reed, voles were often seen, at a frequency of about one in every 50 metres travelled. In some places the burrowing of the voles over the years had caused the canal to widen into the adjacent fields. Some 'bays' created this way are

*Canals prove to be a good stronghold for water voles provided the banks are not reinforced with sheet metal piling or concrete.*

periodically repaired by the canal engineers. This has often led to the widescale use of sheet metal piling, making the bank untenable for water voles. Today, where possible, vole friendly maintenance and repair techniques are being used. On the Oxford canal one such repair involved plastic pipes, of a diameter that voles can go through, being inserted down the burrow entrances to retain a route to the newly repaired canalside. Coconut fibre matting formed the new canal wall and the 'bay' was filled in with clay soil. The voles stayed on site while the work took place and were seen cheekily stealing the coconut fibre for bedding material. The vole colony still survives there today.

**Water Volewatch**

Over the period March to October 1997, Water Volewatch was launched as a public participation survey aimed at the junior branch of The Wildlife Trusts and involving 30,000 young people together with their families, schools and friends. Thanks to sponsorship from Norsk Hydro, some 130,000 user friendly survey forms had been printed and their returns analysed by Oxford and Newcastle Universities. This comprised one of Britain's largest volunteer led mammal surveys ever staged. Although not directed at systematic coverage of the country as in the national survey, the sheer number of people out looking and the far ranging publicity surrounding Water Volewatch was expected to identify a fair proportion of the best water vole sites remaining in Britain (as well as locating isolated populations that may be in need of conservation help).

# How many are there in Britain?

Estimating the total British population size of water voles has been difficult due to their patchy distribution. However, one estimate suggests a pre-breeding (i.e. over wintering) figure of around 1,200,000 animals with a summer population up to 2,500,000. These figures were derived from the relationship of the number of water vole latrines counted over a given length of waterway to the number of occupied territories (from work carried out in Yorkshire by Gordon Woodroffe) and then extrapolated to other river systems throughout the UK. The estimate was only meant to provide a rough guide to how many there are and so assumed that the water voles were evenly distributed along all waterways the same (which is not the case). The frequency at which water voles occurred in the different regions of Britain during the 1989-90 survey was used to refine the estimate but it still could be wildly inaccurate.

Obviously, a more detailed attempt at a total population is now required, since the different rivers can support differing densities of water voles and the relationship between latrine counts and vole numbers may vary considerably in different parts of the country.

Indications from recent survey work suggest that we may now only have one fifth of the sites occupied by voles that were occupied during 1989-90: a very dramatic decline. If this is a true revised estimate of the above, then the new calculated population figures could be as low as 200,000 water voles (or approximately 3,000 colonies) in the whole of Britain and they are still declining. Two hundred thousand may sound a lot when you compare it with some figures for very rare birds that are in the low hundreds, but you must remember that water voles are at the bottom of the food chain. There would have been well over a million colonies at the start of the twentieth century. How does this estimate compare with other widespread species? Figures for other mammals carefully calculated from survey and population data collated at Bristol University by Professor Stephen Harris in a *Review of British mammals* (published 1995) include pre-breeding estimates of 37 million rabbits; 31 million moles; 7 million brown rats; 23 million bank voles and a staggering 75 million field voles. Their estimates for scarcer mammals include 7,000 otters, 100,000 mink and a mere 1,300 black rats.

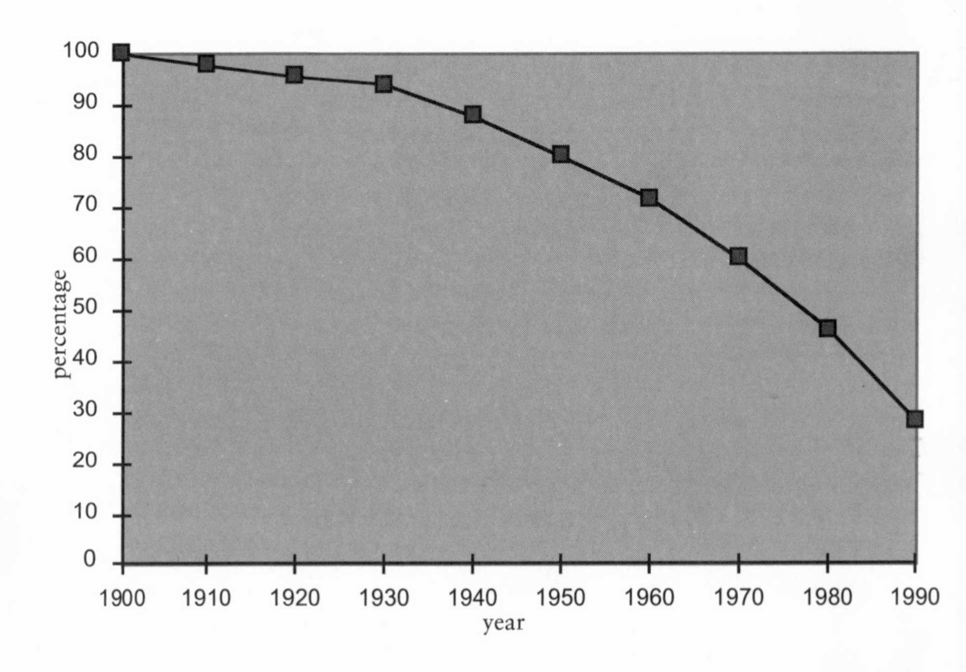


*Change in abundance of water voles in Britain from 1900 to 1990 (derived from the accumulated loss of recorded historical sites re-surveyed in 1990. Source:* The Water Vole in Britain 1989-90, *Vincent Wildlife Trust).*

# How long do they live ?

*Exceptionally water voles may survive three winters but mortality is very high in the first year especially among dispersing animals. Water voles may need to attain a body weight of at least 170g in order for them to survive the winter. One detailed study of water voles where every individual had been captured, marked and then released found that during winter the population may be reduced to 30% of the former summer level.*

# Changing fortunes of Ratty

**Earliest records**
Archaeological excavations at a number of sites dating back to the last glacial and interglacial periods of the middle Pleistocene have yielded teeth and bones of the extinct large vole from which our present water vole probably evolved.

The present day species does not appear until the early Flandrian post-glacial period and the earliest record is from the faunal remains at Nazeing, Essex, dated around 10,500 years before present. There was still a high arctic climate at this time and an ice-sheet remaining in Scotland. At about 10,300 years b.p. a sudden warming of the climate took place and within about 50 years the conditions changed from arctic to warm temperate and there was an associated improvement in vegetation succession. Interestingly, water voles are documented from the fauna at the Mesolithic archeological site of Thatcham, Berkshire (10,050-9,600 years b.p.) but not Star Carr, Yorkshire (9,500 years b.p.). This suggests that the water vole was slow to colonise Britain as it occurred at the older southern site but not the younger more northern site.

This may explain the fact that water voles have never been found on the Scottish islands or Ireland and the Isle of Man. The period of minimum sea level around Scotland was 8,000-7,800 years b.p. offering a land bridge at that time to the above places.

It is apparent from the vast quantities of water vole teeth among the sub-fossil record of bone-cave deposits that the species was extremely abundant in former times. It is interesting to speculate that perhaps, before the arrival by introduction of rabbits and man's domestic livestock, the vole community were the dominant grazers in Britain, especially at upland sites where they may have led a more terrestrial existence. The subsequent history of the water vole in Britain has been largely governed by the activities of man changing and managing the environment.

**1900 – present**
Some indication of how formerly abundant the water vole once was in Britain can be seen by looking at the early writings of naturalists in a series of books published between 1880 and 1920. These are the *Victoria County Histories* and the *County Vertebrate Faunas*, that described the occurrence and status of all the known mammals, birds, reptiles and amphibians for

each county. Typical comments for the water vole record the following:

Cornwall (dated 1906): 'common in all suitable habitats throughout the County'; Devon (1906): 'common everywhere'; Shropshire (1906) 'very common on pools and streams'; Cheshire (1910):'abundant; inhabiting the banks of all unpolluted streams, the meres and smaller pools'; Lakeland (1892): 'common throughout Lakeland from Furness to the Scottish borders, upon which it is especially numerous'; Tweed area (1911) 'any kind of country seems to suit it, so long as there is water at hand. It is equally at home for instance, by the marshes on the coast, the ditches bordering cornfields, the ponds in the midst of plantations or the streams meandering among the hills, it may even reach considerable elevation in the Scottish highlands'.

Up to 1950, written references to water voles are sparse but some indication of relative abundance is provided by the accounts of the trapping campaign to eradicate feral musk rats in the 1930s. (By 1932, musk rats that had escaped from fur farms were breeding in the wild and had inhabited rivers and marshes of 14 counties, the largest concentrations in Shropshire, Sussex and Perthshire). Trapping from 1933-37 accounted for 2,305 water voles accidentally killed instead of musk rats in Scotland alone.

Since 1950, there has been increasing post-war interest in mammals and this is reflected in the surge in recorded localities for the water vole in books and journals. In particular there was a peak in the number of water vole records for the 1960s and 1970s. This was the result of better recording rather than an actual peak in the number of water voles in the country and was largely due to the efforts by The Mammal Society, who at that time aimed to obtain nationwide distribution records for the first British Mammal Atlas, producing much additional field work. However, this still tended to reflect the distribution of observers rather than a true picture of water vole whereabouts.

During the 1980s, anecdotal evidence (by way of local reports and observations) suggested the disappearance of water voles from previously long occupied sites and this prompted a 'desk survey' on the species. This enquiry used an analysis of the data contained in County Mammal Reports and 'old' literature going back to the early 1900s, supplemented by a questionnaire and information from the Waterways Birds Survey of the British Trust for Ornithology. The results indicated that the British water vole population may have suffered a long-term decline in this country since at least 1900.

To follow this up an objective and systematic survey took place over 1989-90 to form the basis for a conservation plan and future monitoring of the water vole's status.

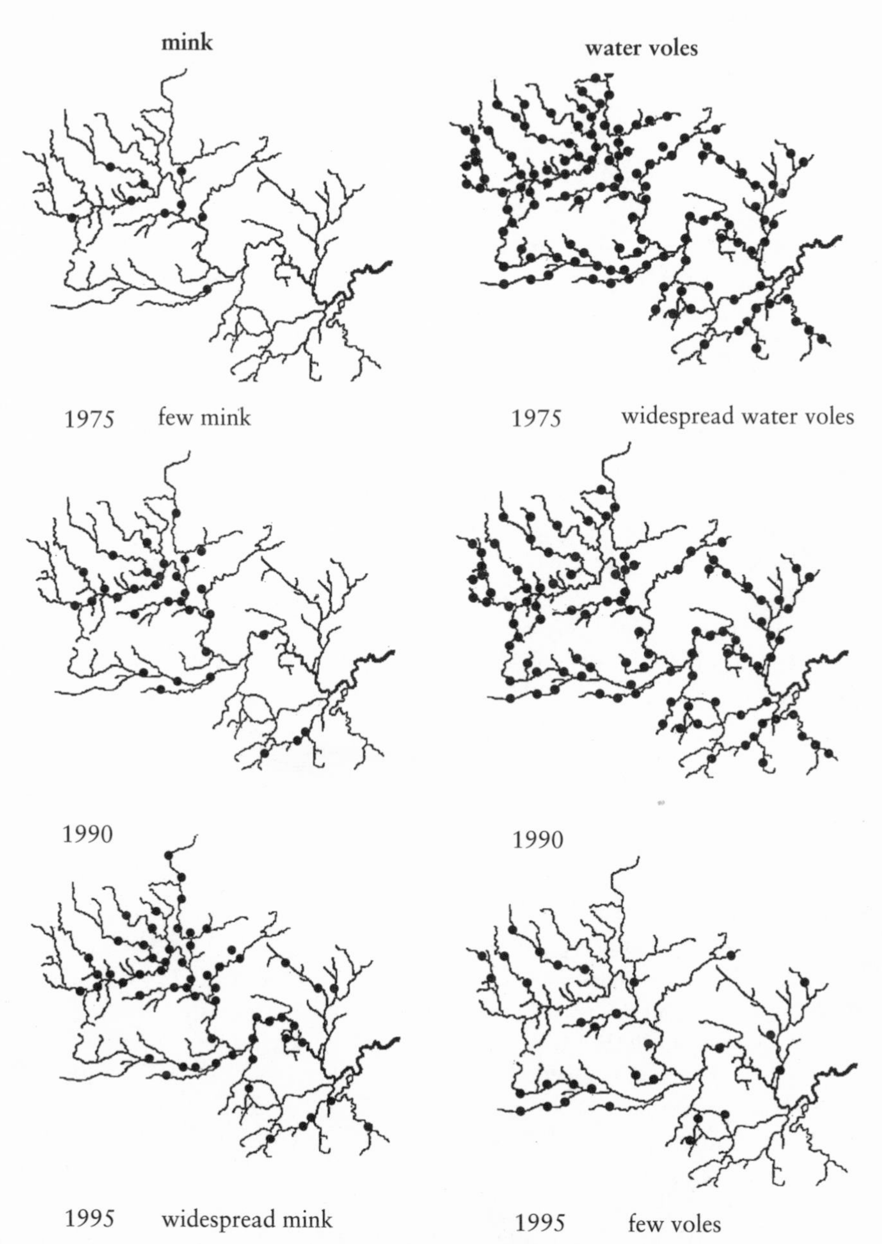


*RIGHT Loss of sites used by water voles along the River Thames 1975, 1990, 1995; LEFT Arrival and spread of mink along the River Thames for the same periods (author's data).*

# Pointing the finger of blame

*'Leaving the water-side, where rushes stood thick and tall in a stream that was becoming sluggish and low, he wandered country-wards, crossed a field or two of pasturage already looking dusty and parched, and thrust into the great realm of wheat, yellow, wavy, and murmurous, full of quiet motion and small whisperings ...*

*Restlessly the Rat wandered off once more, climbed the slope that rose gently from the north bank of the river, and lay looking out towards the great ring of Downs that barred his vision further southwards ...; today, the unseen was everything, the unknown the only real fact of life.'*

Various hypotheses, none of them exclusive, have been proposed to explain the nationwide decline in the water vole.

The factors potentially contributing to the water vole's decline include the degradation or fragmentation of habitat, isolation of water vole colonies, changes in the fluctuation of water levels, pollution, and the impact of predators. Each is a problem in its own right but together the effects may be catastrophic to the survival of water voles.

**Habitat degradation**

In the North York Moors National Park, it was found that highly layered vegetation was characteristic of water vole breeding sites whereas sites with poor cover yielded little or no evidence of the species. Factors implicated in the degradation of riparian habitat include:

**(i) Heavy grazing pressure by domestic livestock**

Recent investigations have demonstrated that the more grazing there is on the riverbank, the less likely the presence of water voles. Heavily grazed sites tend not only to be denuded of luxuriant growth, but also to be heavily poached.

**(ii) River engineering and bankside maintenance works**

Attempts to improve land drainage by dredging, canalisation and clearance of bankside vegetation and trees have increased considerably during the second half of this century.

It seems obvious that river engineering and bank maintenance, by removing food and cover, will be detrimental to water voles in the short-term; but need not have long-term impact provided there are suitable sites for water

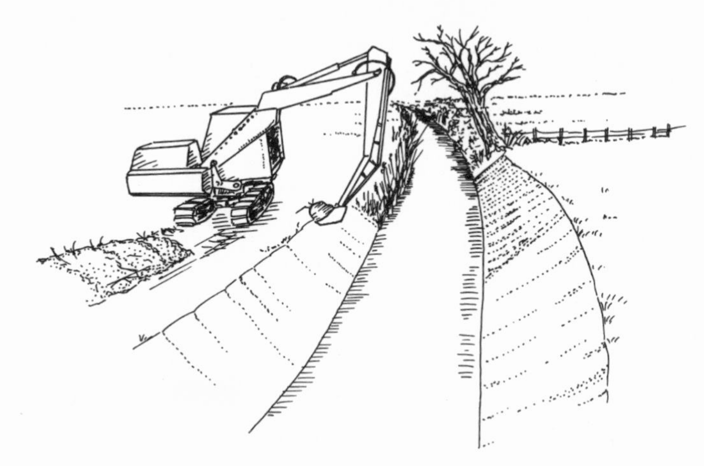

*River engineering work, desilting and re-profiling the river channel can have devastating consequences for water voles.*

voles to retreat to nearby. One study demonstrated that the dredging and re-sloping of a drainage dyke in Lancashire displaced the water vole population for over 6 months and it then took a further 12 months for the numbers to reach former levels, during which time the vegetation had again become dense. Riparian engineering works are known to severely reduce the numbers of breeding birds present on waterways, especially moorhen and coot. Heavily managed sections contain fewer territories, birds breed later and produce fewer second clutches. As water voles, like moorhens, make extensive use of riverside plants that are removed by such works, we might expect them to be similarly disadvantaged. Irrespective of any recovery that bankside and aquatic vegetation may make after riparian engineering, one goal of such work is to increase water flow, and this is likely to lead to greater and more sudden fluctuations in water levels.

**(iii) Bank reinforcement**

Recreational power boating causes a wash which undercuts and erodes banks, inhibiting plant growth and exposing and damaging tunnels. Plants are damaged at mooring sites and also by oil pollution. Reinforcement of the riverbank against the effect of boatwash erosion has largely been unsympathetic to the needs of water voles, involving the use of sheet piling or concrete blocks.

**Habitat fragmentation**

Habitat fragmentation, resulting in smaller segments of suitable habitats separated by greater distances, is a universal hazard to species conservation. First principles dictate that at some level of fragmentation, loss of suitable bankside vegetation will disrupt the population dynamics of water voles. However, in the absence of much basic biological information about these rodents (for example, dispersal distances and the details of their mating system) it is hard to predict the consequences of particular patterns of habitat fragmentation. But what is certain, the more isolated a colony becomes the less likely it will survive in the long term.

**Fluctuations in water level**

Access by water voles to food, cover and burrows along riverbanks will be affected by fluctuations in water level. The pattern and extent of such fluctuations might be affected by changes in land use and associated land drainage, by climatic change and by flood control policy.

**(i) Changes in land-use and associated land drainage**

Using aerial photographs, a study at Oxford University revealed substantial changes in land-use throughout the last half century along the floodplain habitat of the Thames valley. In particular, the coverage of semi-natural grassland fell from 49% of the whole area in 1947 to 12% in 1991; there was a concomitant rise in tilled acreage from 11% to 53%.

The effect of post-war agricultural intensification on wildlife has involved the loss of many valuable farmland habitats. For instance, the loss of hedgerows (and often their associated ditches) may impede water vole dispersal, whereas the replacement of semi-natural grassland by tilled fields involves not only a wider reduction in biodiversity but also removes from the floodplain a 'natural sponge'. Tilled land is efficiently drained, and so we may expect the observed changes in land use to be reflected in changed patterns of water flow and flooding.

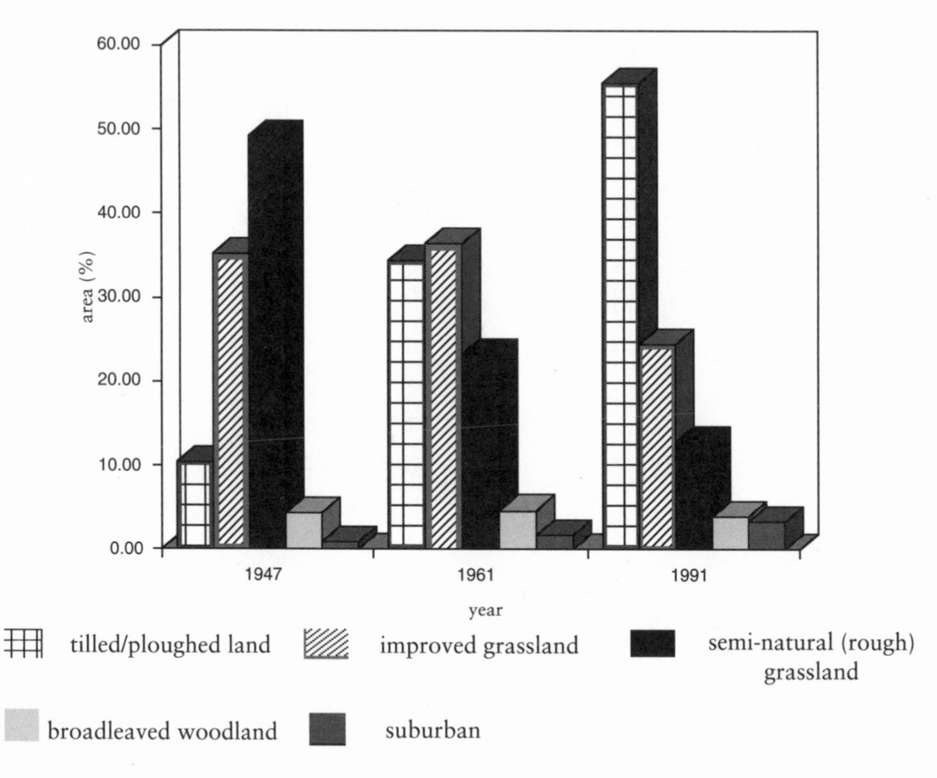


*Change in land-use in the River Thames floodplain, determined from a series of aerial photographs taken in 1947, 1961 and 1991. The sequence shows a loss of semi-natural floodplain grassland in favour of improved grass and then tilled (and extensively drained) land. Field boundary hedgerows and ditches have also been lost as small fields have been converted to much larger ones.*

## (ii) Flooding and climatic change

The combined effects of land-use changes and short- or long-term variation in rainfall influence the flow of water through rivers and the pattern of flooding. A recent investigation demonstrated that throughout the last 55 years within the Thames catchment there has been great variation in both total annual rainfall, maximum flood levels and the number of days on which the river floods.

Low rainfall periods combined with abstraction for crop irrigation has led to low river flows and drought conditions. Many springs completely run dry, leaving no water in ditches, ponds or feeder streams to rivers. Water voles need water – as a means of escaping predators and also as a quick way of patrolling their territories. Drought conditions may also parch the riverbank and reduce the essential food supply and cover in which to hide. Severe weather might affect water vole lives in various ways. Populations have been shown to be reduced after prolonged periods of flooding or big freezes. Frozen rivers and thick ice may limit access to food and water and the intermittent severe weather might affect water voles in various ways. The prolonged presence of thick ice may limit access to food and water, and the intermittent occurrence of harsh winters (e.g. 1916-17, 1939-40, 1946-47 and 1962-63) is similarly detrimental to birds requiring access to open water, such as herons and kingfisher. However, the most obvious climatic danger to water voles is flooding. As the river rises the water voles seek refuge in air pockets within their tunnel systems, but when the water brims to the bank or floods over it for several days the voles are forced away into unfamiliar habitat. Obviously this happened in the past too, but the reclamation of the floodplain land for agriculture has meant that the number of sanctuaries they can rely on in times of flooding have diminished. It seems likely that this increases mortality, due to predation, hunger and chilling associated with any flood event. A succession of such events, by chance or as part of a climatic trend, as evidenced in the early 1990s, would worsen this mortality.

**Pollution**

Contaminants of the freshwater and riparian environments which could conceivably affect water voles include organochlorine insecticides and their metabolites, polychlorinated biphenyls, heavy metals, and organic pollution from farmyard fertilizers, slurry and sewage. Organochlorine insecticides and polychlorinated biphenyls are found in freshwater systems throughout Britain and, being lipid soluble, pass along food chains in animal material. Consequently, they are found as contaminants in freshwater fish, birds such as herons and otters. Although they are known to produce many detrimental sub-lethal effects, it seems unlikely that herbivorous water voles will accumulate high residues.

DDE and certain alkyl phenols (such as nonyl phenol) widely used in industrial processes have oestrogenic affects on mammals (that is, they mimic the female hormone, oestrogen, and so reduce the fertility of males), and have been found in rivers below waste water outfalls, but their effects on water voles remain unknown. Perhaps the water vole is acting like a coal miner's canary, warning us that we should urgently study these pollutants in our aquatic ecosystem. However, it is clear that many rivers, like the Thames, were formerly even more polluted than they are today,

being completely de-oxygenated during the nineteenth century through the release of untreated sewage from the growing human population. Water quality generally improved over the 1970s and 1980s, although the level of organic pollution may have worsened again since 1985 due to sewage pollution, farm waste, run-off from farm land and drought conditions. In short, while the hypothesis cannot be excluded, we have not yet done the research to show that pollution is a major cause of the decline in water voles.

**Disease**

At present, there is no evidence for or against the hypothesis that any disease has contributed to the water voles' decline. This is largely due to the lack of research directed at wild water vole populations. The occurrence and effect of Weil's Disease (Leptospirosis) and other diseases prevalent in brown rat populations and their possible transmission to water voles are also unknown.

**Competition**

Competition with brown rats for space and food has been suggested as a possible factor in the water vole's decline when the habitat deteriorates for them but its effects are presently unknown. Water voles do seem to avoid rats (see later) and there have been recorded instances of vole colonies being taken over by brown rats, especially in some urban situations. Where rats and water voles occur together, the use of rodenticides to control rat infestations may also cause problems by poisoning the voles accidentally.

# Vole au vent

*'Weasels – and stoats – and foxes – and so on ... Well, you can't really trust them, and that's the fact.'*

In Britain and Continental Europe the water vole has had to contend with a wide range of traditional predators for many millennia, including fox, otter, stoat, weasel, rat, owl, heron, raptors and large fish. Typically water voles constitute less than 5% of the predator's diet, although each individual predator may catch a larger proportion locally at certain times of the year depending on vole abundance. Among the avian predators, water voles can be regularly found in barn owl and short-eared owl pellets with local studies showing them comprising 30% of the diet. Astonishingly it is the most important prey item of the hawk owl and ural owl of Northern Europe forming over 90% of their diets. Pellet studies can however be misleading, over-representing the mammals in the diet. This is the case with herons which eject vole pellets regularly but not fish bones which are more easily digested. Obviously fish constitute the heron's staple diet.

The Scottish naturalist, Dick Balharry, observed that water voles regularly turned up in the prey items being brought to golden eagle eyries in the western Highlands of Scotland. This confirms the presence of water voles at remote upland sites and presumably open areas that form the hunting territory of such golden eagles. Water voles were also reported taken by golden eagles among the flow-country of Caithness and Sutherland.

Among the mammalian predators in Britain, three species have given cause for concern as to their effect on water vole populations. Firstly, the **brown rat**, which has been reported to predate on nestling and juvenile water voles and even cause local extinctions or displacements. During the 1970s and 1980s, 'plagues of rats' were reported in a widespread number of local newspapers throughout Britain and were reported to have been particularly numerous along many waterways.

Investigating an urban population of water voles on the banks of the River Wear, in Durham City, David Knight found that the nighttime activity of water voles became reduced when brown rats spread into the study area. The voles were found to spend a longer period of time foraging in the late afternoon and evening but then did not emerge from their burrows until after sunrise. Before the rats had arrived at the site, the voles could be

*Barn owls are supreme hunters of water voles.*

found being active for short periods throughout the night at roughly three hourly intervals. This suggests that the water voles were actively avoiding any encounters with rats, which have been known to hunt and kill them.

Brown rats have been recorded taking over the burrow systems of water voles, enlarging the tunnels, digging down to the nests to predate on the nestlings and, it is said, causing small local populations of voles to go extinct.

Yet it is possible that brown rats do not always get it their own way. Stephanie Ryder records an experiment where she introduced a brown rat in a cage trap into her vole pen, whereapon all members of the water vole family who were old enough to fight immediately gathered around the trap and tried to chase the rat away. Another instance of the water vole's ability to defend itself was reported to me by a pest control officer who used a multi-doored cage trap to clear a rat infestation. This allows him to cage whole families of rats at the same time. When he went to check the next morning he found five rats cowering in one corner and a big male water vole facing them at the other side of the trap, apparently unscathed.

Secondly, **domestic cats** have also been implicated in local heavy predation pressure, with individual cats repeatedly hunting a small patch so successfully as to cause local extinctions.

More recently the arrival of the feral **American mink** has been implicated in

the water vole decline. The impact of this species is discussed in detail below.

**Survival strategies**

As a prey species the water vole has evolved a number of anti-predator strategies:

(i) **Hide or escape to safety**

(a) The first option is to sit still and avoid being seen; this works with birds of prey – owls and herons that hunt by sight. It does not work with predators that hunt by smell.

(b) The next option is to escape to the burrow system, especially 'bolt holes'. This works with all predators except those slim enough to follow them underground, such as stoats and weasels.

(c) The final option is to dive into the water. Voles purposefully splash dive in a noisy 'plop' that acts as a warning signal to other voles. The vole then kicks up a cloud of sediment to act as a screen to confuse predators.

(ii) **High population recruitment:** as in many prey species, water voles show r-selected characters; that is, small body size, early reproduction, quick growth and large numbers of young. So, although short lived, they have between 2-5 litters a year and each litter is between 2-6 young. Young born before July may breed in the same year although most do not reach sexual maturity until after their first winter (adults rarely survive 3 winters). Predators usually account for surplus animals in the colony.

(iii) **High recolonisation potential:** among a local population there is a natural dispersal of juveniles together with an emigration of adults defeated in territorial disputes, including some pregnant females. Thus predation losses at some sites may be quickly replaced by immigration.

# Nasty smell

*Recent research is suggesting that the sense of smell is very important to all rodents and that the odours of predators may cause the prey species to avoid an area. The passage along a riverbank by stoats, weasels and mink may cause the water voles to hide, move away or even panic, disrupting their normal behaviour. So even if these predators are not hunting the voles themselves their very presence may upset the social system and perhaps the breeding success at the site.*

# Predation by American mink

There has been much past debate between conservationists on the issue of the extent to which introduced American mink have been a problem to our native fauna. But hard facts in the debate have been lacking. However, more recent studies have indicated strongly a connection between the loss of water vole sites and the pressure of mink. For instance, Gordon Woodroffe in the North York National Park was present when one tagged water vole was predated by a mink, which he first saw searching all the burrow entrances before locating the vole in dense cover and killing it after it had dived in the water.

The evasive behaviour of kicking up sediment to act as a screen has been seen to be completely ineffective against mink. Also, adult voles appear to be close to the optimum prey size for adult mink, which means they make a one-day food package all in one. In North America the muskrat (*Ondatra zibethicus*), which is slightly larger than the water vole, is perhaps the mink's most important mammal prey.

Along the River Soar in Leicestershire, Chris Strachan was further able to show that vole populations within the territories of breeding female mink can be decimated within only one year. The demands of rearing kits mean that the female mink hunts intensively close to the nursery den and in so doing not only can find the local vole colonies but probably every vole. She is also slim enough to follow the vole underground.

When American mink colonise a river they do so, initially, in selected habitat types, leaving the voles to occupy the intervening riverbank between these mink areas. This system suggests a further, and very serious impediment to water vole dispersal: the journey between different colonies of voles may be through territory heavily hunted by mink.

**History of mink in Britain**

American mink were first imported into Britain from Canada and Alaska in the late 1920s to be farmed for fur; from then until 1945 the industry was a small one. The business expanded after the war and by 1962 the number of mink keepers had risen to a peak of around 700 with an annual pelt production of 160,000. In 1962, the provisions of the Destructive Imported Animals Act 1932 were expanded to include mink, by the Mink (Keeping) Order. This tightened the conditions under which mink were kept, to stop them escaping as well as improving their husbandry requirements.

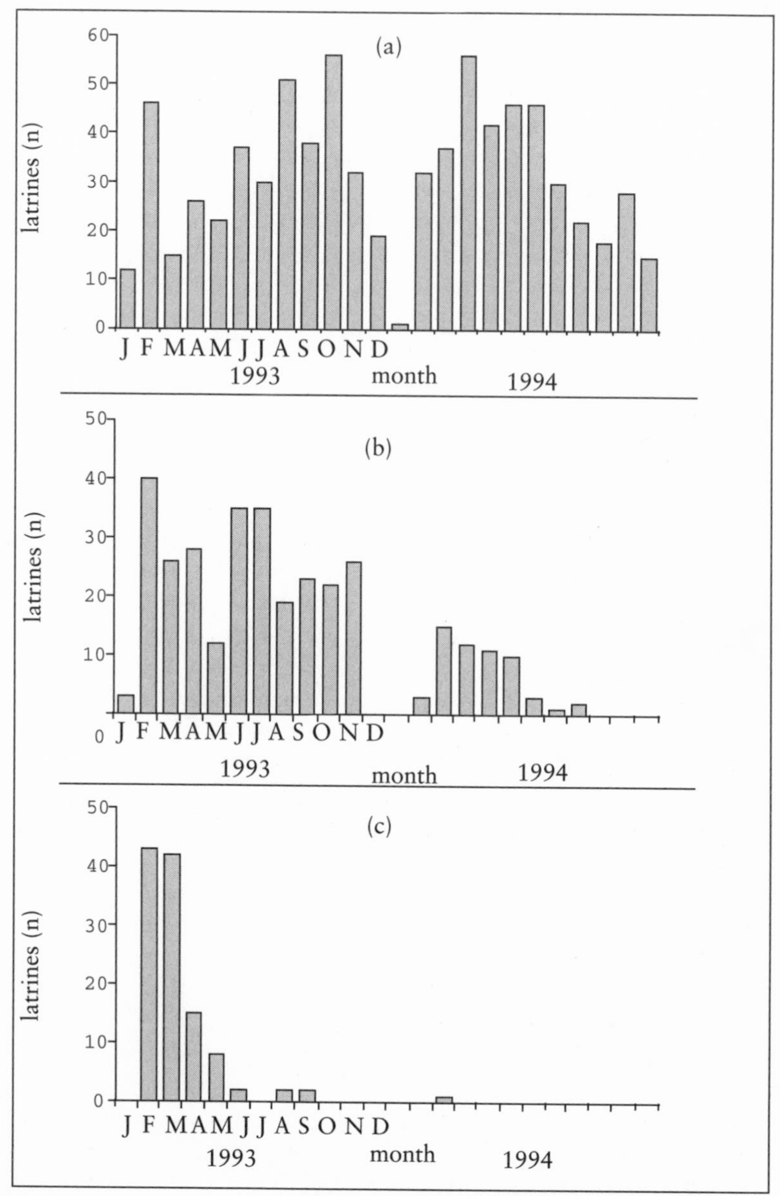


*Monthly pattern of water vole latrines (1993-4) suggesting the relative strength of the vole colony at (a) River Amber (no mink present), (b) River Soar (mink arrived winter 1993/4) and (c) River Soar (mink arrived winter 1992/3)*

*In Britain water voles have faced a whole range of predators for many centuries.*

The number of farms dwindled to about 240 by 1971 (although the production of pelts had risen to around 300,000) and by 1984 there were only 54 farms still operating. In the 1930s some of the mink had escaped from captivity but breeding in the wild was not confirmed until 1957 when a female with young was seen on the River Teign in Devon. This proved a

*The arrival of mink in Britain has led to water voles facing a new predator never experienced before.*

nucleus for subsequent colonisation of neighbouring river systems.

Similar populations became established in the early 1960s in those areas liberally supplied with mink farms, namely Hampshire, Wiltshire, Sussex, Lancashire, West Yorkshire, South Wales, Aberdeenshire and Perthshire. At the same time, efforts to control feral mink were stepped up by the Ministry of Agriculture, Fisheries and Food (MAFF) and by the mid 1970s, 4,875 mink had been caught from 41 counties in England and Wales and a further 1,946 from 29 counties of Scotland.

During the 1970s MAFF decided to cease their efforts at control as it had proved ineffective and costly. Nevertheless thousands of mink continue to be trapped and shot in Britain each year by gamekeepers, farmers, water bailiffs, angling clubs and owners of wildfowl collections. However, total eradication from an area has been shown to be difficult to achieve and then the resulting vacuum is only temporary.

The continued spread of the mink in England, Scotland and Wales from 1977 to 1994 has been mapped by the three sets of National Otter Surveys. But perhaps the most detailed known distribution, that for 1989-90, was obtained during the national water vole survey, being repeated in 1997.

**Can mink and water voles co-exist?**
Of course there are many sites where mink and water voles can and do co-exist side by side. Overall, the 1989-90 survey recorded co-existence at 17.6% of all the sites occupied by water voles, and it is encouraging to note that where water voles remain in those regions of high mink density (such as south- west, north-west and Northumbrian regions) up to 37.5% of those sites showed co-existence. This may be partly due to the fact that mink have just moved into these sites and so reduction or elimination of water voles may follow. However, at others they have been co-existing for many years and this may be due to particular habitats. The places where they have co-existed for some time typically possess extensive areas of dense and luxuriant vegetation cover or are adjacent to high quality wetland such as willow carr or reed fen. Examples of such sites include Stodmarsh in Kent, where mink and water voles have co-existed for over 20 years and Tregaron Bog on the upper Teifi in Wales, where the population of water voles survives alongside mink, polecat, otter and avian predators.

On some river catchments (notably the spate rivers of upland Britain) the 1989-90 survey located water voles mainly at the head-waters, on side streams, ditches and brooks away from the main rivers and large tributaries. It is likely that these smaller water courses are peripheral to the mink's home range (the core of which occupies the main river and larger tributaries) and so are hunted less frequently. The local distribution of the water vole has been reduced by predation so that the two species are eventually separated into differing parts of the same area with the mink taking its preferred habitat and the water vole moving into the other areas.

*Cors Caron Nature Reserve (Tregaron Bog), where mink and water voles have co-existed for over twenty years.*

# Ratty on a tightrope

Loss of good riverbank habitat through changes in agricultural land-use and improved land drainage, the canalising of rivers and the artificial reinforcement of banks to speed water flow and prevent flooding have resulted in the direct loss of vole habitat and the fragmentation of the vole colonies.

Pollution and disease may have played a role in the decline of water voles in the UK, but the arrival of the mink on the waterways appears to have had a significant impact on the water vole and may be the last straw for its ultimate survival.

The inter-relationship between all these factors has been summed up in 'The Tightrope Hypothesis' – that agricultural intensification has confined the water vole to a narrow ribbon of riparian habitat and in so doing has increased its vulnerability to mink predation.

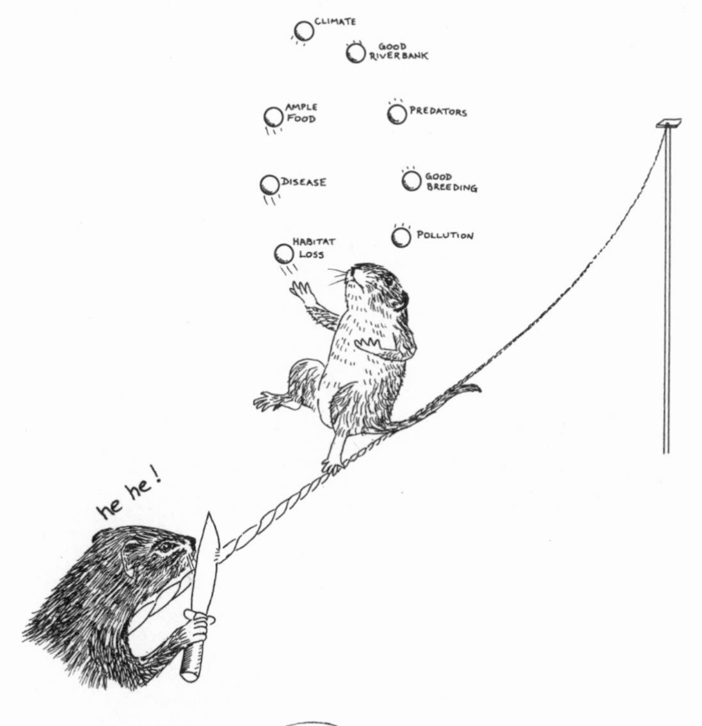

Studies into whether this is the case are now being done; if they show that habitat improvement would help voles significantly, it is not too late to do something about it.

The hypothesis predicts that American mink and water voles could co-exist if the water voles were freed from the linearising constraint of a narrow ribbon of bankside vegetation. Both can be found living side by side in areas of irregularly shaped expanses of highly structured wetland. The 'unimproved' natural floodplains of eastern Europe provide a natural example in contrast with the British situation: in Belarus, for example, along rivers swathed in many metres of lush reeds water voles apparently co-exist with a community of semi-aquatic predators that include not only American mink, but also European mink, polecats and otters.

A different balance of factors may be involved in the decline of the water vole in other regions of Britain. For example, areas with faster flowing rivers may naturally have supported fewer voles and offered circumstances where they were more exposed to predation. If so, then in these areas the mink may more profoundly be the root of the problem.

American mink arrived in the UK at a time when polecats and otters were rare. Perhaps the recovery of these species will affect mink numbers; perhaps, also, if these native predators had been more abundant they too, even in the absence of American mink, would have been toppling water voles off their tightrope of habitat during the twentieth century. These possibilities are testable and are currently being tested. They also form an important part of designing a water vole recovery plan.

In conclusion, of course it would have been better if the American mink had stayed in America, but it is probably here to stay. At least in lowland catchments such as  the Thames valley, it appears that the mink's impact on voles could only occur because the mink arrived in the wake of human destruction of riparian habitats. If so, then while intensive mink control may be an effective management option for particular conservation areas, it is unlikely to lead to either the eradication of the mink or the salvation of the water vole. Attention might more fruitfully be focused, more cheaply and with many more value added benefits, on the re-creation of riparian habitats within the agricultural landscape.

A vendetta against the mink distracts attention from the wider issues about the health of the riparian environment. This is dangerous for the water vole and many other wetland species besides, by distracting us from the way we manage wetlands and water.

# Practical conservation

Ideally this section should be written after 10 years of solid practical and scientific research into the exacting habitat requirements, population processes and tried and tested conservation methods for preserving the species. However, to do nothing until we have such information may mean we are too late, so what follows has been produced as an interim guide to the best conservation and habitat management options we have to date. Many of the suggested management guidelines are as yet untested but are considered beneficial on current knowledge.

They will be reviewed and revised in the light of forthcoming research results. One thing is certain, water voles need help and they need it now!

A number of practical conservation measures are suggested below to maintain existing water vole populations and encourage the restoration of other populations that have declined locally.

**Protecting existing habitat**

Reed beds, sedge beds, stands of emergent vegetation, riverbank tall herb and tall grasslands are essential for a self sustaining water vole population. Excessive over-grazing by sheep, cattle and horses not only reduces the amount of food and cover for the voles but the poaching of the ground at the water's edge makes the site untenable by compacting the soil and damaging the burrow system. At such sites it is recommended that the bank is fenced or at least part fenced to provide refuges for the water voles.

Fencing off meanders may be an effective way of allowing natural regeneration and retention of water vole habitat, using only a small amount of fencing materials (consideration must be made for livestock access to the river, angling access and flood damage).

Temporary exclusion from the watercourse is possible through the use of electrified fencing. The fencing is movable so differing regimes/ widths of protected areas could be managed and grazing animals could be allowed periodic access to prevent the dominance of coarse, rank and scrubby vegetation taking over the bank.

I saw a good example of sympathetic fencing on the River Coln, at Stowell Park, Gloucestershire (in the Upper Thames catchment). After erecting electrified fencing along both banks of the River Coln, with cattle bays to allow cattle to drink, emergent and mid-channel regrowth was quickly re-colonised by water voles from an upstream source. Excessive

grazing and bank poaching had previously made the site untenable for them. This section of the River Coln had been targetted as a brown trout spawning ground and the exclusion of cattle was essential to ensure the survival of the fishery as well as the water voles.

If mowing of the riverbanks has to be done, areas of uncut vegetation can be left, perhaps as patches of varying length (depending on what the site is like) at close intervals. This can be done on opposite banks or one bank can be left uncut. By altering the cutting regime annually, woody scrub will be prevented but sufficient food and cover will be left for water voles. A 2 metre band of bank cover in patches of 20-50 metres would be ideal.

Unsympathetic weedcutting can pose a serious problem to water voles by removing cover and food resources over very large areas; in the Northern Area of the Anglian Region of the Environment Agency over 1,500km of river are weedcut each year, which could cause very harmful results for water voles. Typically, weedcutting is carried out over several kilometres in just a few days, resulting in the complete removal of all the growth of the waterside plants. Here, sensitive management options would be to retain marginal vegetation along one or both banks in 50m strips or to cut opposite banks in alternate years.

Where river channel management is required, dredging from one bank only is recommended, in line with current good practice. In most instances, it should be possible to leave stands and/or marginal fringes of waterside vegetation during de-silting operations in order to retain valuable habitat for water voles. This is particularly vital where there are voles present and there are no other nearby water features which may act as refuges. A method of clearing channels and leaving shelving along banks may be a suitable option for encouraging the regrowth of a fringe of tall emergent vegetation.

River maintenance operations should normally leave banks untouched, but where bank re-grading is thought necessary, this should be of a profile which maximises the width of marginal vegetation that can re-establish itself. The bank profile could be stepped, or with a steeper incline on the upper half of the bank, to facilitate burrowing, Any bank re-grading should be restricted to small sections as much as possible; it is preferable to retain existing bank profiles, particularly if well vegetated. Nearby waterways or lateral channels should be left untouched as refuges for the water voles from which they can recolonise these temporarily disturbed habitats once the bankside vegetation has regenerated.

Particular care should be taken when excavating and re-dredging adjacent ditches, small streams and lateral channels as these are important features for the local water vole population. A survey for water vole signs prior to

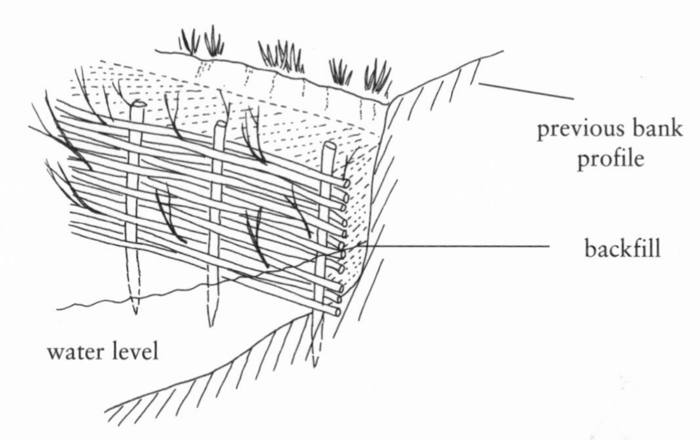


*Where riverbanks need protection from erosion the use of willow poles woven with living branches (withies) may provide an option that will sprout and strike roots into the bank to provide a living and permanent protection that is friendly to water voles.*

work may be necessary to establish the extent and core of the local population, and as much as possible of the fringing vegetation should be retained at the core.

Where bank erosion requires bank reinforcement, the use of sheet metal piling, rock gabions or masonry should be avoided at water vole sites. Sympathetic bank maintenance should be encouraged by way of small scale repairs or through the use of living willow withies, coir fibre bundles or other natural materials which will allow the bank to be used by water voles following repair.

**Water level management**

Both excessive flooding and excessive drying of a site can make it untenable for water voles. Long-term stability of water levels appears to be an important prerequisite for their survival at a site. The practice of lowering water levels in rivers and drainage systems during winter is detrimental to aquatic margins, particularly during dry winters when they may be exposed to severe frosting.

The introduction of sluices and weirs regulates natural processes in rivers and is generally regarded as environmentally degrading. However, the upstream retention of water and control of water level has been shown to have great benefit for water voles by creating ponded habitat with features similar to backwaters (i.e. luxuriant marginal and in-channel vegetation growth). Minor sluices and weirs can enhance degraded in-channel habitats

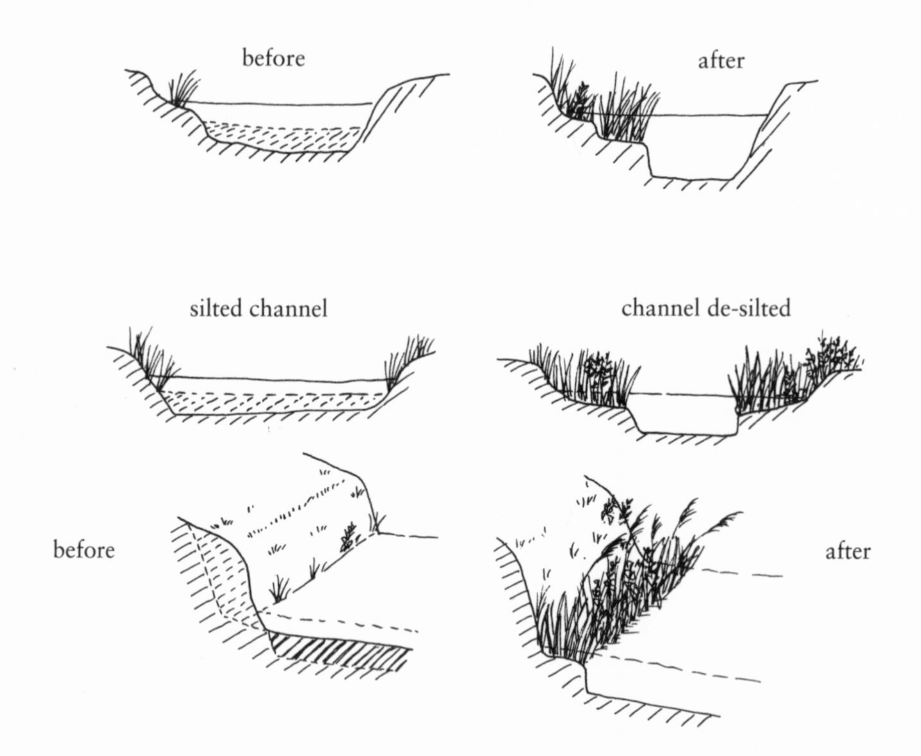


*Sympathetic channel deepening or widening to create shelves of extensive water plant fringes may be essential habitat enhancement work for water voles.*

and floodplain wetlands, especially by diverting flows into secondary channels or ditches.

A good example of where weirs have helped retain water voles is on the River Derwent, at Chatsworth, Derbyshire. The slow flowing stable water levels above drop weirs provide conditions for large water vole colonies that are largely absent from elsewhere on the watercourse.

**Habitat enhancement**

Habitat can be enhanced for water voles by the creation of suitable banks in which to burrow and the establishment of a broad band of emergent and bankside cover. At its simplest this may just involve fencing and allowing the vegetation to regenerate itself naturally. Bank re-grading or two stage channelling may be helped by the planting of suitable species.

Opposite is a table of species that may be considered as important for water voles providing both food and cover. Seed mixes of these could be sown in any restoration or habitat creation projects.

**Grasses**
*Phalaris arundinacea, Phragmites australis, Lolium perenne, Poa trivalis, Poa pratensis, Dactylis glomerata, Glyceria maxima, G.fluitans, G.plicata, Arrhenatherum elatius, Deschampsia cespitosa, Anthoxanthum odoratum, Holcus lanatus, H. mollis, Agrostis stolonifera, Phleum pratense, Alopecurus geniculatus, Alopecurus pratensis, Molinia caerulea*

**Rushes**
*Juncus inflexus, J. effusus, J. conglomeratus, J. acutiflorus,J. articulatus*

**Sedges**
*Carex paniculata, C. diandra, C. otrubae, C. hirta, C. riparia, C. rostrata, C. vesicaria, C. pendula, C. laevigata*

**Water weeds**
*Sparganium erectum, S.emersum, Sagittaria sagittaifolia, Alisma plantago-aquatica, Butomus umbellatus, Potamogeton natans, Ceratophyllum demersum, Myriophyllum spicatum, M.verticillatum, Iris pseudacorus, Menyanthes trifoliata, Nymphoides peltata, Nuphar lutea, Nymphaea alba*

**Herbaceous plants**
*Polygonum amphibium, Caltha palustris, Ranunculus sceleratus, R.flammula, R.lingua, R.peltatus, R.aquatilis, R.fluitans, Alliaria petiolata, Nasturtium officinale, Cardamine amara, C.pratensis, Filipendula ulmaria, Geum rivale, Potentilla palustris, Lythrum salicaria, Chamerion angustifolium, Epilobium hirsutum, Anthriscus sylvestris, Myrrhis odorata, Aegopodium podagraria, Oenanthe aquatica, Apium graveolens, A.nodiflorum, Angelica sylvestris, Galium palustre, Myosotis scorpioides, Mentha aquatica, Veronica beccabunga, Valeriana officinalis, Sonchus palustris, Taraxacum officinale.*

Ponds, old oxbows, backwater channels and a network of floodplain ditches are extremely valuable for water voles and should be re-instated if possible (such as where marginal farmland permits them). The restoration of water meadows at Sherborne Park Estate (National Trust), Gloucestershire, is an excellent example of what can be done. Traditionally, water meadows were created to produce a large area of good quality early grass for the grazing of sheep and cattle, through the intentional flooding of riverside fields by using a series of sluices, carriers and drains. At Sherborne such meadows were created in 1844, but these fell into disuse in the early 1930s due to changes in farming practices. The grassland was then improved by ploughing, re-seeding and applied with artificial fertilisers. From 1965 the land was used for cereal production until 1992 when, under the ownership and management of the National Trust, the land was put into Countryside Stewardship Scheme to restore 57 hectares back to a water meadows system. 10,000 metres of ditches were re-dug using original plans and former aerial photographs; the ditches were to carry water onto the meadows from the River Windrush and back into the main channel again. Original stone sluices that had remained intact but buried in the meadows were located and restored and the whole site was re-fenced to control grazing pressure (8,000 metres of fencing).

The restoration of Sherborne water meadows is a good example of countryside stewardship in practice, creating excellent water vole habitat. We wait with bated breath for the voles to discover it from the nearest colony 5 km upstream. Although deliberately flooded in winter, the water levels are controlled and maintained by sluices so that the voles should have non-flooded channels in which to move and benefit from the rich feeding provided by lush vegetation.

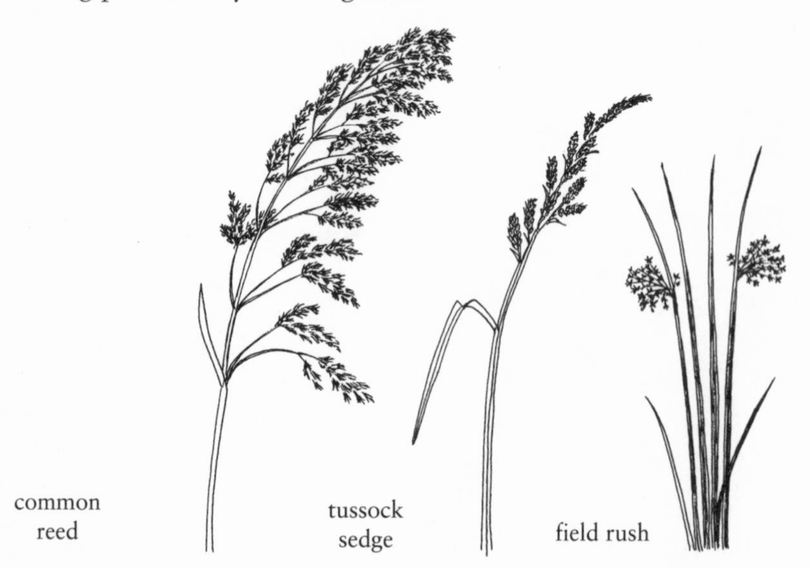


*Plants that provide good food and cover are essential for voles.*

*Protecting the water margin from excessive trampling and over grazing by domestic animals can easily be achieved by fencing.*

Creation of new wetlands or restoration of traditional ones with the features I have mentioned is desirable, and care should be taken with the site design to maximise the amount of riparian frontage (overall bank length) so as to provide sufficient area that can support a self-sustaining water vole population.

A rule of thumb guide is that one breeding territory requires 50 metres of waterway bank (with female voles using both banks on waterways less than 2 metres wide). A minimal over-wintering population of female water voles may be as low as ten individuals. The overall length of suitable waterway, showing a dense fringe of emergent vegetation should not be less than 500 metres and preferably in excess of 1000 metres. The greater the amount and length of suitable habitat, the greater the possibility of maintaining a viable population of water voles.

**Agri-environment incentive schemes:** A number of options are now available for landowners and farmers to enhance the watercourse habitat on their land for the benefit of wildlife. A number of these schemes offer excellent opportunities to improve habitat for water voles. Present schemes include Countryside Stewardship, Habitat Scheme; Riparian Setaside; Conservation Headlands; Riparian Buffer Zones and a water fringe option within Environmentally Sensitive Areas. These various schemes are entirely voluntary, but incentive payments to take land out of food production for 5 or 10 years offer some compensation to farmers. For more information, contact your local MAFF office.

Water-vole friendly management may be essential to the species survival in some areas of Britain.

# A brighter future for Ratty?

*' ... suddenly he stood by the edge of a full-fed river. Never in his life had he seen a river before – this sleek, sinuous, full-bodied animal, chasing and chuckling, gripping things with a gurgle and leaving them with a laugh, to fling itself on fresh playmates that shook themselves free, and were caught and held again. All was a-shake and a-shiver – glints and gleams and sparkles, rustle and swirl, chatter and bubble. The Mole was bewitched, entranced, fascinated.*

*'By the side of the river he trotted as one trots, when very small, by the side of a man who holds one spellbound by exciting stories; and when tired at last, he sat on the bank, while the river still chattered on to him, a babbling procession of the best stories in the world, sent from the heart of the earth to be told at last to the insatiable sea.'*

What does the future hold for the water vole? What stories will the babbling river have to tell about Ratty's fate fifty years from now? One thing is certain: that if we do not act now then the sight a water vole swimming buoyantly across the water surface, the sound of a 'plop' as one dives under and the bright twinkle of an eye from within a snug, cosy burrow, will not be there for future generations to enjoy.

In 1996 the water vole was included in a list of nine British Mammals for a Biodiversity Steering Group Report. This has lead to a special working group of Government organisations and scientists being set up to draw up and implement a Species Action Plan for the water vole.

In 1997 the water vole was added to the special protection schedule of the Wildlife and Countryside Act, making it an offence to destroy the places where water voles live.

Legal protection that is aimed at individual animals from acts such as persecution, exploitation and disturbance at places of refuge is not, on its own, sufficient to arrest the decline of the water vole. These threats are probably not major ones that jeopardise the population as a whole. However, far more serious is damage and destruction of riparian habitat that results in fragmentation and loss of water vole populations, as this is one of the major underlying threats facing the species. Legal protection afforded under Schedule 5 of the Wildlife & Countryside Act 1981 (as amended 1997) ensures that the presence of water voles at a site must be taken into account by riparian habitat managers when carrying out engineering and maintenance operations and by planning authorities when considering development proposals. Environmental impact assessments are pre-requisites for all new housing and road schemes: these look for the presence of water voles (as well as otters, great crested newts and other scheduled species).

Provisions in the Act allow for injured or disabled animals to be taken and kept for the purpose of looking after them; similarly, badly injured animals may be killed as an act of mercy. The law also permits actions which would otherwise be illegal, provided these are the incidental result of a lawful operation and could not reasonably have been avoided.

Statutory Nature Conservation Organisations can issue licences to allow otherwise prohibited actions (such as trapping and handling) for reasons of conservation, scientific research, education, photography and other legitimate activities.

Advice must be sought and licences granted before any potentially damaging operation that may effect water voles is carried out.

Addresses of the SNCOs for England, Scotland and Wales are given in Useful Addresses.

This is only a general guide to the main provisions of the law; for more detailed information The Wildlife and Countryside Act 1981 (as amended 1997), should be consulted.

**Species action plan**

During 1997 an Action Plan was produced by the UK Biodiversity Steering Group for the water vole to provide guidance for the species conservation. Its long-term aims are:

- To arrest the decline and maintain the current distribution and status of the water vole in Britain.
- To restore water voles to their former widespread distribution (i.e. pre 1970s) by the year 2010.
- To ensure management of watercourse and wetland habitat which will maintain the restored population.

Specific objectives include:

1. To incorporate water vole conservation into national and regional policy.
2. To maintain the current distribution and abundance of the species in

the UK.

3. To promote water-vole friendly methods of riparian and watercourse management.

4. To restore areas of habitat within the former range in order to support expansion of the current population.

5. To encourage the re-establishment of water voles at restored sites and aim to ensure that they are present throughout their 1970s range by the year 2010.

6. To continue the necessary conservation related research and population monitoring.

7. To use the water vole as a flagship species for good riparian and wetland habitat.

Water vole conservation is beneficial to a diverse array of plants and animals. The Environment Agency is the contact point for the Species Action Plan and a Water Vole Habitat Management and Conservation Handbook providing technical and practical advice is due to be published by the end of 1997.

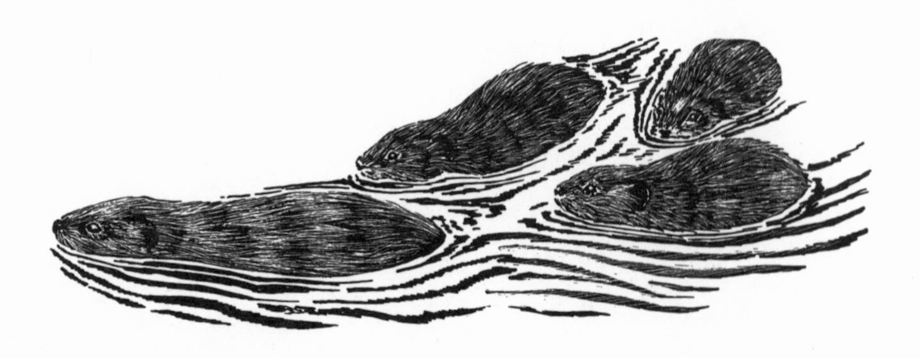

*The sight of families of water voles on our waterways can be ensured for the future.*

# Vole sanctuaries

County Wildlife Trusts should perhaps consider 'water vole reserves' in those areas where water vole populations are still strong and suitable management can be carried out on a local scale.

Also, water vole populations at the various Wildfowl and Wetland Trust centres (such as Slimbridge, Llanelli and Martin Mere) have found a relatively safe sanctuary from predators (apart from the occasional heron or owl) and benefit from the provision of grain used to feed wildfowl in the collection pens. WWT centres should be recognised as having high conservation value in safeguarding their water vole populations in the face of the serious declines in the wider countryside.

1996 saw the first scientific study of the water vole at Slimbridge, unprecedented in the history of research carried out by an organisation primarily concerned with birds. In collaboration with Oxford University, the study focused on monitoring the numbers, distribution and trends at sites within and outside the predator-proof fence, quantifying home ranges and basic population parameters, including recruitment, survival, immigration and emigration for the Slimbridge population.

All valuable background information from which to draw up good management guidelines for the species.

# Reintroduction and translocation

Re-population by releases of voles captured in similar local habitats would be feasible if areas were created or re-created for them with their habitat requirements and causes of decline borne in mind. This would need to be where later protection could be afforded to them.

Studies of many mammal species, such as otters, badgers and dormice have shown that if they are simply let go in an unfamiliar place, they will scatter and probably most will die. This may not only be cruel but is wasteful of resources. Any release scheme needs careful planning, guidance and support.

**Stage 1**
Site review: Plans for any release must have reviewed the site for the following points (in accordance to IUCN guidelines):

- There should be good historical evidence of previous occurrence at or near the site.
- Natural recolonisation is thought to be unlikely.
- The factors causing local extinction should have been identified and rectified.
- Available habitat must be sufficiently extensive to support a viable population.
- Genetic provenance of the release stock should be considered with the aim being to introduce animals from populations that originated as close to the release site as possible.
- Obtaining animals for release should not jeopardise existing wild populations.

**Stage 2**
Consultation: It is important to ensure agreement in advance with landowners before the release and about possible modification to existing management plans (sites should be carefully chosen considering habitat requirements for the water voles, together with current and future threats to the animals and their habitat). Nature organisations must be consulted in relation to any reintroductions involving Sites of Special Scientific Interest.

**Stage 3**
Pre-release preparation: Site management (suitable food and cover), release

cages should be built so that water voles are allowed to become accustomed to a site prior to release.

Size of release group: water voles are noted for their aggressive behaviour to each other so each cage should contain a social group that have been living together (male and female/ mother and young) or just single animals. Cages could be sited at 100 metre intervals. Timing of release is also important and should be when vegetation is growing (spring/early summer).

**Stage 4**
The final stage of any release programmes is the careful monitoring of its success or failure.

# Captive breeding

*A potential 'eleventh hour' option for the species is captive breeding for release back into the wild. If this is to work successfully then we will need to establish co-ordinated breeding centres of local provenance. One such potential centre is Nature Quest in the New Forest where water voles are already being bred for their excellent riverbank display of British wild mammals.*

*The captive breeding of water voles has been shown to be extremely difficult under laboratory conditions. This is largely due to the need for each animal to have a large space in the enclosure. Fighting is very common between individuals and if more than one female and one male are put together, animals may get killed. Outdoor enclosures that provide an area for burrowing, underground nest chambers and a fresh water pool in which the voles can plunge bathe, are much more successful. The floor space of such enclosures needs to be at least 2m x 2m in dimension (and preferably larger). Galvanised metal sheets make good enclosure walls that the voles can not climb up (wire mesh walls not only need to be of a small hole size, about 1cm square, but need to have a wide over-hanging lip of sheet metal to stop the voles escaping) and each enclosure needs to have a wire mesh floor below the ground to stop the voles tunnelling out.*

*If the young are not removed from the cage before the adult female has given birth to a second litter, they will be attacked and even killed by the mother. Once breeding has begun, a single female may be encouraged to produce four or five litters over the course of the season. This way, in just one year up to 100 young could be produced by a small original stock of perhaps just 2 males and 4 females.*

# Epitaph

In conclusion it is worrying that a previously common mammal, such as the water vole, should have declined to such a degree over such an extended period without being noticed by the numerous watchful professional and amateur naturalists and conservationists. Local concerns were first being voiced only a decade ago and continuing apparent losses suggested the need of legal protection and a specially prepared Species Action Plan document.

One final point: the unnoticed decline of the water vole suggests that one could have widespread and catastrophic declines in populations of other less visible common mammals, such as pygmy or common shrews, remaining undetected for many more decades. Only if the species was an important prey of predatory birds (e.g. field vole) would the decline be detected because of a secondary reduction in the numbers of the more noticeable predator (e.g. as when the common buzzard declined after the decimation of the rabbit population due to myxomatosis). This emphasizes the importance of the long-term surveys of small mammal populations at fixed stations distributed over Britain, and indicates that more studies should be carried out on changes in status of those mammals which are presently considered widespread and abundant.

# Useful information

## Addresses

Water Vole Steering Group:
Lead Agency – Environment Agency
Environment Agency – Thames
Kings Meadow House
Kings Meadow Road
Reading RG1 8DQ

Scottish Environment Protection Agency
SEPA
Irskine Court
Castle Business Park
Stirling
SK9 4TR

**Statutory Nature Conservation Organisations**
For England: English Nature
Northminster House
Peterborough
PE1 1UA

For Scotland: Scottish Natural Heritage
2 Andersen Place
Edinburgh
Scotland EH6 5NP

For Wales: Countryside Council for Wales
Plas Penrhos
Ffordd
Penrhos
Bangor
Gwynedd LL57 2LQ

Water Vole Technical Support Group
WildCRU
Dept. Zoology
University of Oxford
South Parks Road
Oxford OX1 3PS

British Waterways
Environmental and Scientific Services
Llanthony Warehouse
Gloucester Docks
Gloucester GL1 2BJ

The Mammal Society
15 Cloisters Business Centre
8 Battersea Park Road
London SW8 4BG

The Wildlife Trusts
The Green
Witham Park
Waterside South
Lincoln LN5 7JR

## Further reading

If you would like to delve more into the lives of water voles, then there are surprisingly few accounts specifically written on the subject. However, a couple of booklets do spring to mind; the first is no longer in print but you might find a copy in a second hand book shop – *Water Voles: Animals of Britain No.4* by Stephanie Ryder (Sunday Times Publication, 1962), this has some excellent black and white photographs and firsthand observations by the author; the second has just been produced by the Mammal Society (1996) – *The water vole* by Gordon Woodroffe: this is a useful introduction to the subject and has some good colour photos.

Other useful accounts that draw on scientific publications are found in the *Handbook of British Mammals 3rd edition*, edited by Corbet and Harris (Blackwell, 1991) and The Vincent Wildlife Trust survey report published in 1993, called *The Water Vole in Britain 1989-90: Its Distribution and Changing Status* by Rob Strachan and Don Jefferies. The latter not only shows the most up to date distribution maps but extensively reviews the literature.

For practical advice on how to conserve water voles and manage their habitat a Water Vole Conservation Manual will be produced by the Environment Agency autumn 1997.

# Index

## British Natural History Series